Future Telecommunications

Information Applications, Services, & Infrastructure

Future Telecommunications

Information Applications, Services, & Infrastructure

Robert K. Heldman

With contributions by

T. F. Madison

T. A. Bystrzycki

McGraw-Hill, Inc.

Washington, D.C. New York San Francisco Blue Ridge Summit, Pa. Auckland
Bogotá Caracas Lisbon London Madrid Mexico City Montreal Milan
New Delhi San Juan Singapore Sydney Tokyo Toronto

Library of Congress Cataloging-in-Publication Data

Heldman, Robert K.
    Future telecommunications / Robert K. Heldman.
        p.   cm.
    Includes index.
    ISBN 0-07-028039-8 (h)
        1. Telecommunication—Technological innovations.   2. Information
    technology.   3. Information services.   I. Title.
    HE7631.H45   1992
    302.2—dc20   92-21416

    5 6 7 8 9 0   DOC/DOC   9 6 5 4

ISBN 0-07-028039-8

*The sponsoring editor was Neil Levine, the editor for this book was Sally
Anne Glover, the designer was Jaclyn J. Boone, and the production supervisor
was Katherine G. Brown. This book was set in Century Oldstyle.*

Printed and bound by R.R. Donnelley & Sons Company.

To Valerie, Pat, and Linda.
To our children,
our children's children,
and the future children
of the forthcoming
Global Information Society.

Contents

Part I
The information millennium

Part II
Information applications

Part III
The emerging global information society

Part IV
The information decade

Prologue
A new frontier

It is . . .
A new time,
A time of expansion
A time of new thinking,
A new society,
A Global Information Society,
A new age,
The Age of Information,
A new millennium . . .
The Information Millennium . . .

It is indeed a new frontier—a time to use the exciting advances in technology to address the many needs of today's society. It's time to turn our attention to peaceful applications of technology that advance and elevate our human existence. It's time for courageous leadership to explore the unknown. Now is the time to enter this new frontier, with its challenging opportunities. It's time for information—unlimited information—to be accessed and processed to help us establish a new society—a global society—a Global Information Society—in the Age of Information—the Information Millennium.

Thomas A. Bystrzycki
Vice President
Network & Technology Services
U S WEST Communications

Irwin Dorros
Executive Vice President
Bellcore

A. Hasholzner
CEO and President
Siemens Stromberg-Carlson

K. W. Hovaldt
Senior Vice President
Fujitsu Network Switching of America, Inc.

Leif Kallen
President and CEO
Ericsson North America Inc.

Thomas F. Madison
President, Markets
U S WEST Communications

John S. Mayo
President
AT&T Bell Laboratories

Roy Merrills
Senior Vice President and President
Northern Telecom–United States

David Orr
President and CEO
Alcatel Network Systems, Inc.

Foreword
A vision of the future

Tom Madison
Tom Bystrzycki

The structure had no shape,
representing a shapeless form,
yet to be determined,
in a future yet to be . . .

As we look to the future in terms of its technical possibilities and market opportunities, we can't ignore the magnitude of change that will be forthcoming, affecting all nations, all lands. With each passing day, we're faced with new challenges, new decisions, as we pursue the vision "to be the best in the world in connecting people, the generators and users of information, with their world."

During these turbulent times, it's essential to have a clear understanding of where we're going and how we'll get there. Without the long view, our focus on everyday needs can become so near term, so intense, that we overreact to less important aspects and miss the essential issues that have long-term impact. As we address the changing needs of the information marketplace, we need to understand the complete application from the customer's perspective. Each task, each function, each operation must be visualized in terms of using not only today's telecommunications information tools, but also tomorrow's. We must encourage and aid our customers to grow in their use of communications and meet their needs by providing timely, flexible, controllable, economical, and available services within easy access and selection.

To do so, we must formulate a marketing view that encompasses the full range of information services. It's no longer a voice-only world. We must make room not only for data and text, but also for image, graphic, and video information. We must use today's existing plant to its fullest in order to provide as many new information services as possible. However, while enhancing the current plant to provide these new services, we must, at the same time, deploy new fiber for the new, exciting, visual services for the turn of the century and centuries to come, as the Information Millennium draws nearer.

With this in mind, each industry, each market sector, each application must be considered in terms of the full range of narrowband, wideband, and broadband technical possibilities. Once new services are identified, to achieve them, we must provide a plan of action that can be realistically implemented—for today's limited resources must be used to maximize their impact in resolving the greatest range of needs. This can only be accomplished with a realistic vision that provides the most comprehensive view that can be determined at this point in time.

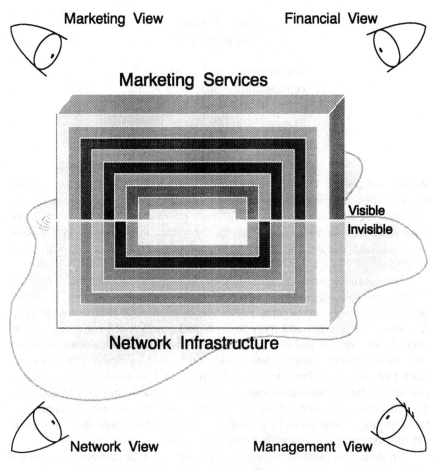

Fig. F-1. Information infrastructure.

As indicated in FIG. F-1, we not only need to visualize the specific information services that we want to deliver, but we also need to identify the somewhat invisible network structures that support these services; it's ineffective to finally decide what services are needed by the information users, but then not have a network that can enable such offerings for many years to come. New services need new networks; new networks need a sufficiently long lead time to be deployed, and the switching systems for new networks need an even longer lead time to be researched, designed, tested, and installed.

In order to successfully meet the needs of the computer industry, the communications industry must provide a public information-handling infrastructure that enables private-to-public internetworking. It must facilitate rapid delivery of new services by multiple vendors, but it must do so in a manner that maintains the basic network transport integrity. Accomplishing this requires an understanding, by all parties, of the necessary steps to obtaining the functional infrastructure—from both the service and the network perspective. Understanding and acceptance can only be reached if each step in achieving the desired infrastructure is properly phased to ensure that financial cost considerations and service-pricing incentives are satisfied in order to enable sufficient customer acceptance and use, thereby meeting the revenue objectives of the stockholders and shareholders. In this manner, each step becomes a building block, helping to generate the needed revenues from which to take the next step—to help launch the next array of services, as we move from narrowband to wideband to broadband, ubiquitous voice, data, and video services.

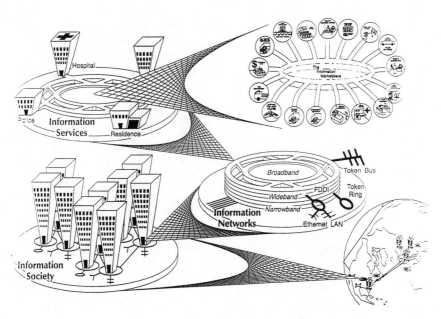

Fig. F-2. Global information society.

Once the complete picture becomes evident, as all the pieces of the puzzle begin to come together, then the enhanced service providers, alternative network providers, state and federal regulators and overseers, as well as the switching, transport, and customer-premise equipment suppliers, the providers, and their customers will all be able to better appreciate each others' roles as we work together to restructure America's telecommunications infrastructure.

The following analysis will help us all step back and take the long view, the broad view, from both a technological and a service perspective (see FIG. F-2). We need to review the marketplace in terms of specific applications, then consider the steps we might want to take together to achieve the necessary narrowband-wideband-broadband telecommunications infrastructure to make America a formidable and competitive participant in the forthcoming Global Information Society.

> *"Each one of us adds a little*
> *to our understanding of nature,*
> *and from all the facts assembled*
> *arises a certain grandeur."*
>
> **Aristotle**

Acknowledgments

This book is for all within the industry who have, over the years, contributed to the advancement of communications and computer technologies and their applications to better serve our human needs. This is presently a time of great controversy and complexity, as we seek to build upon these past achievements and go forward with greater expectations of a better world to be. With this in mind, I hope this analysis will help establish a plausible direction for our endeavors. I've taken considerable care to provide just the right information in the right form for your consideration—not too much, not too little. I hope you find these thoughts beneficial to your thinking; I hope they "tickle your mind" with new ideas and possibilities for meeting the challenges and opportunities of the forthcoming Information Millennium.

It's been an exciting time working with John Willemssen, Tom Madison, and Tom Bystrzycki, as well as my many friends and associates within U S WEST, GTE, ITT, AT&T, Bell Labs, Bell Communications Research, Siemens, Northern Telecom, Fujitsu, NEC, Ericsson, Digital, IBM, North American Rockwell, and various research institutes and universities. Special thanks to my editors, Larry Hager, Neil Levine, and Sally Glover of McGraw-Hill, and to my son, Peter, whose research assistance, analyses, and reviews have greatly helped to properly focus this series on the appropriate issues, challenges, and opportunities. Also, special thanks to my wife, Valerie, and my daughter, Catherine, and my many secretaries—Becky, Carol, JoAnn, Livya, and Shirley. During the past 30 years or so, they have transcribed my scribbles into legible thoughts.

Introduction
The third millennium

The triumph of democracy . . .
"And even should the cloud of barbarism and despotism again obscure the science and libraries of Europe, this country remains to preserve and restore light and liberty to them. In short, the flames kindled on the fourth of July, 1776, have spread over too much of the globe to be extinguished by the feeble engines of despotism; on the contrary, they will consume these engines and all who work them."

Thomas Jefferson

As we climb to the pinnacle of our local mountain and peer into the vast wilderness of a previously hidden range of mountains, we can't help but be exhilarated by what lies beyond, with all its challenges and opportunities. In pausing to look back from where we came, before stepping forward, it's interesting to note how the past century, the 20th century, has become the portal to not only a new millennium but also to a new era, the Age of Information, which brings with it a new society—the Global Information Society. As nations change and past barriers rapidly disappear, the third millennium becomes the formation period of a new global village, with all its hopes and opportunities, as it addresses many of the key, complex riddles of the somewhat elusive human's search for a better quality of life.

This new age, the Information Age, brings with it new complexities and choices, but this is the game of life—to search, pursue, and achieve as we attempt to climb each step of Maslow's hierarchy; first we strive to satisfy basic survival needs of food, clothing, and housing; then we pursue health, security, wealth, power, work satisfaction, respect, and then on to the personal satisfaction of self-fulfillment and achievements.

So, what is the third millennium? What are its challenges and opportunities?

Indeed, we'll most likely find life quite human, quite personal, quite individual, but interrelated, quite globally interdependent. The past two-thousand years of western Christendom have uplifted humans from enslaving each other. World wars of the 20th century have brought an end to the class consciousness remaining from the Victorian society of upstairs/downstairs living, where domestic servants lived and worked downstairs, under the house, for minimal wages. Postwar American technology has been transferred around the world, along with democratic beliefs, to provide a continuously exciting flow of new machines that ease household chores and free both sexes to pursue success, as the definition of "success" changes, in a increasingly complex marketplace.

Japanese, Asian, and European firms have quickly adapted and advanced American management techniques to competitively harness and channel changing technologies into new inventions to meet the transportation and communications needs of the rising, new, young consumer group, whose more expensive education and advanced trade and office skills enable them to demand higher incomes than their parents. However, as technology and its complexities advance, there appears a widening gap between the haves and the have-nots within the industrial and the not-so-industrial nations. We've seen the collapse of the USSR into a commonwealth of multiple republics, each locked in its own desperate struggle for survival, as it attempts to "catch up" with its Western counterparts. We've also seen the United States and its allies' superior technology (with Japanese help) defeat Iraq's armies in the Gulf war. This has caused would-be dictators to scramble to maintain or regain power and obtain this technology (as well as the new nuclear technology and the scientists left over from the former Soviet Union). The Cold War might be over, but the global economic wars are just beginning. The Japanese have clearly established themselves as the initial victor on this front, with their major advances and achievements in high-quality, high-tech products.

This then sets the stage for the third millennium, as U.S. debt approaches 4 trillion dollars and economic hard times are challenging the world communities, where numerous countries face recessions, dissolving currencies, less demand for their raw materials, growing populations, growing unemployment, and intense competition from a more technically complex and market-wise, competitive global community.

As the world population stretches to 6 billion, as technological excellence further separates the more industrial countries from the rural Third World, as years of industrial pollution and unrestrained waste severely tax the fragile ecology, while the continuing loss of moral values, drug-related crime, and diminishing family units cause rising human suffering and despair, we, the collective human race, timidly enter a new millennium, with both its opportunities and challenges.

As we enter this new era, we must better deploy our technologies to help everyone, the growing 6 to 7 billion of us, live more successfully together, thereby peacefully obtaining a higher quality of life globally. We must use and replenish our planet's resources in a universe that a being far more intelligent than ourselves, whom we call God, has provided for us. Once successfully achieved, these accomplishments will enable us, in time, to explore and settle the other planets of our fascinating galaxy, and, in future milleniums, to explore and settle other, previously hidden galaxies. As NASA targets colonization of Mars at the end of the 21st century, one of our major challenges on planet Earth is to effectively use computers and communications to establish a global village, one in which members can obtain a higher quality of life by teaming together and using information technology to solve the problems created by an expanding population, a global village in which personal aims and expectations are unbound. . . So it was; so it is; so it could be . . .

*"Seize the future
because, . . .
it's yours!"*
John F. Kennedy

Part I

The information millennium
A new frontier

Information technology
Information networks
Information applications
Information services
Information users
Information society

1

Society's Technology— technology's society

"So, here we are, committed by our Greek origins to a life of asking questions, that provide answers, that create more questions with no end in sight. As our amazing abilities become even more amazing by the more questions we ask, are we now reaching the stage that it is no longer a question of what novelty and change the future will bring next, but what kind of future we care to invent—make happen, because we cannot leave well enough alone . . . ?"

Burke

As we near the end of the second millennium, we see that society is indeed, as Burke noted, an evolutionary, technology-based entity, taking many twists and turns as it matures and develops. To better understand where we're going, let's take a look at where we've been and how we got here. In so doing, we'll see how today's society is really a result of yesterday's technology, and how today's technology will create tomorrow's society . . .

Over the course of human events we have indeed asked many questions to which we did and did not get correct answers. Many times, political restrictions inhibited the pursuit of knowledge, as resistance was given to answers that challenged pre-established views. This was the lot of the wide-scope thinkers such as Aristotle (384–322 B.C.). Later, some church leaders felt threatened when conflict arose in their interpretation of biblical events, such as the manner in which humans were created or whether the Earth or Sun was the center of the galaxy/universe. There was also a time of lack of questioning, causing a lack of change and growth. Religious leaders such as Augustine reacted to the excesses of Roman life by focusing primarily on spiritual life after death, with limited concern for the current problems of human existence.

This caused a total withdrawal of human advancement. The religious leaders cloistered poetry, grammar, writing, arithmetic, and geometry in the monasteries, removing this knowledge from the general public, who were then left with only the elders' memories to recount facts and incidents. (See FIG. 1-1.)

But this is getting ahead of the story; in order to better appreciate the key shifts in technology that helped form today's society, let's take an in-depth look at the advancement of civilization in terms of the society-technology-society-technology cycle that enabled mankind to develop over the past several million years. In so doing, we should also note that if technology had not constantly progressed and expanded to change our way of life, if these technologies had not been available, their absence would have formed a quite different society than we have today.

The first innovations appeared during the Paleolithic period or Old Stone Age from approximately 2,600,000 B.C. to 800 B.C.. During the Paleolithic period, people learned to make axes, chisels, and bows and arrows. At the end of this period, they invented the hoe, sickle, and other tools for growing better crops. As early as 3500 B.C., people learned to melt tin and copper and mix them together to produce bronze—a stronger and durable material. This became the Bronze Age and Iron Age. The wheel appeared around 3000 B.C.; before that, bundles were dragged on heavy sleds called sledges.

During this period in the Nile Valley of Egypt, between the Tigris and Euphrates rivers of Mesopotamia, the people used a vast system of canals, dikes, and ditches to control the flow of water. Later, the Greek inventor Archimedes (200 B.C.), developed the Archimedean Screw to raise water from one

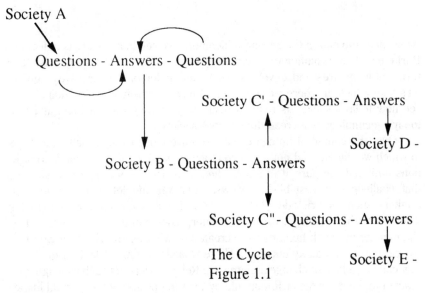

Fig. 1-1. The cycle.

level to another. It's important to note that around 3000 B.C., south of the valley, the Sumerians developed writing as the primary method of remembering knowledge.

As time progressed, China invented paper in 105. At the same time, the Greek Hero was developing the screw press to make wine and olive oil. It should be noted that much was invented in China, but it took a long time for Europe to obtain these inventions, or similar ones. For example, the compass came to Europe around 1100 A.D., and although Pi Sheng made a movable-type printing press in China in 1045, it was not to appear in Europe until 400 years later as the Gutenberg press, which produced the Gutenberg Bible in 1454.

So, civilization progressed over the B.C. period by learning to control fire, cultivate crops (8000 B.C.), domesticate animals, establish permanent dwellings (3500 B.C.), discover the wheel, use metals, develop writings and establish governmental order for the public welfare. In this manner we progressed through the Aegean (3000 B.C.) to the Greek period, which reached its height in 400 B.C., to the Roman rule of 400 B.C. to 476 A.D.. Here, the Greeks wrote with a true alphabet that was derived from the Egyptians through the Phoenicians; the Greeks also provided the first works of history and science. In addition to their architecture and sculpture, they provided the world with the concepts of democracy and justice. Roman achievements included the construction of paved roads, aqueducts and the codification of a system of laws still used today by many nations. These advances and classical learning were preserved by the Byzantine Empire for a thousand years and spread to Central Europe and Eastern Europe. With Christ's coming came Christendom, as Christian beliefs were established across Europe.

During the first millennium A.D., the first 500 years of European society were in the hands of the Romans; after that, history entered the Dark Ages of cloistered learning, with little scientific achievements. During this period, Islamic nations embraced and expanded Greek thinking to include Islamic mathematical and scientific progress. It was not until the Christian crusaders, led by El Cid, took from the Moors control over lower Spain, including the great library of Toledo (1085), that Europeans were awakened by the vast array of knowledge from previous works. These were originally translated to Arabic by Jews and Christians, and were then retranslated to Spanish, then to Latin, for all of Europe to try to understand. Try is the key word, for it was not until Aristotle's works were translated that people were able to index and cross index the information to appreciate how the great works interrelated. In this manner, European thinking was awakened to math, algebra, medicine, biology, zoology, optics, pharmacy, chemistry, astronomy, physics, metallurgy, rhetoric, and architecture, using Aristotle's logic, procedures, and theories to put it all in the proper perspective.

"So we moved from a society based upon the views of the cloistered few to one in which everyone and anyone could say 'give me the facts and

I will think about it'—to make tomorrow better than it was yesterday."

Burke

Thus, we entered the age of transition, preservation, and completion. Here, geometry with proofs was integrated with algebra without proofs to enable, in the late 1600s, Fermat's analytical geometry and its natural consequence, Newton's calculus. This union of two mental outlooks—proof and symbolic manipulation—forms the basis of what a later generation will call "modern mathematics." The crusaders of the 1100s and 1200s introduced many more Islamic skills and learning, "where the light of a new dawn brought an end to the dark ages"[1] . . . This was the destiny of the second millennium.

The 1200s

As the Islamic-Greek works were translated to Latin and studied throughout Europe, the age of logic was established. Unfortunately, the sacking of Constantinople by the Crusaders delivered the death blow to Greek cultures, with the burning of their libraries complete. Crusaders, expecting to meet barbarians, found a culture superior to their own in the Moslem world. Europe then awakened to the challenge. During this period, the University of Paris became the model of all Northern European universities. A later migration of English scholars from Paris established Oxford, and a migration from Oxford began Cambridge. Albert Magnus and Thomas Aquinas became Dominican friars, where Albert, a foremost Aristotelian scholar, pursued physics and astronomy, while Aquinas, his student, later applied Aristotelian logic to provide the philosophical backbone to the Christian faith. As Scholasticism was established, ancient logic and Christian theology were amalgamated.

During the Pao^3Yu^4 Era (1253–8) in China, $Chin^2Chiu^3Shao^2$ wrote his great treatise "Shu Shu Chiu Chang."

Its nine sections of mathematics showed how to solve numerical polynomial equations of varying degrees. Some believe that spectacles were invented around 1285, at the time that TaunLin completed his 348-book encyclopedia, and Kublai Khan, a Lamaist and Confucian, ordered the burning of all Taoist scriptures, save the *Tao Te Ching*. By 1295, the bankers of Florence had been forbidden to use Hindu numbers, while in Baghdad the writings of Greek mathematicians continued to be translated into Arabic. In Spain, as noted, Arabic was being translated into Latin, but in neither case were Greek or Arabic literature included.

---

[1] *History of Mathematics.* IBM. 1966.

The 1300s

During the 1300s, we saw Marco Polo returning to Venice, Dante's writings, and Giotto's painting as universities were being formed in Pisa (1343), Prague (1347), Pavia (1361), Heidelberg (1385), Vienna (1365), and Cracou (1364) and Florence (1335).

However, the Black Plague took its toll as it traveled through India (1332), Russia (1341), and Venice (1348) and then retreated to Russia in 1352 after ravaging Europe. As cannons and gunpowder arrived from China, the literary and research world gathered in the universities to share ideas and wait for what was later to be a key invention for fostering thinking and understanding. (See FIG. 1-2.)

The 1400s

During the 1400s, the Gutenberg press was the long-awaited invention that transformed Europe overnight. In the forty years after its introduction in 1454, it printed more than 8 million books, enabling the foundation of private libraries and private thinking. During this period, Copernicus studied the celestial bodies and turned the world upside down with his observations that the Earth is not the center of the universe. As the Renaissance man was born, it was also the period when one's perspectives were enhanced by one's perception. The Florentine age of perspective was established, in which "one draws what one sees." (Perspective points were established on drawings using focal-point lines to correctly show the scale of all aspects of the building, person or scene.) Next, coordinates became established as "everything is located in its place." This included maps of the known lands and territories, noting the possible shortest distances to the east by circling the maps to show how one could go out across an ocean, rather than go long distances over land to go east. This thinking got Columbus off to discover America in 1492, as he sailed west to go east to the East Indies.

Through the printing press, everyone began sharing ideas, using facts, expanding facts, challenging facts. Now poetry became poetic, as it no longer needed to be simple rhymes in order to enable items or events to be remembered. This brought an end to the need for older people's memories to help establish when and where such an event took place. By the end of this period, weights and measures were becoming uniform throughout Europe, as was thinking, as thoughts and ideas were becoming shared and focused on numerous areas. Books were becoming readily accessible to show these theories and agreements.

"Finally we shall place the sun itself at the center of the Universe. All this is suggested by the systematic profession of events and the har-

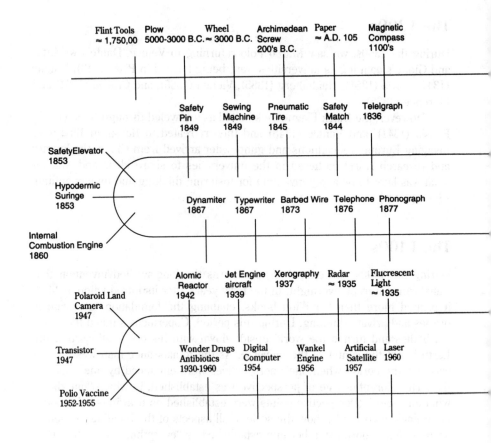

Flint Tools
≈ 1,750,00

Plow
5000-3000 B.C.

Wheel
≈ 3000 B.C.

Archimedean
Screw
200's B.C.

Paper
≈ A.D. 105

Magnetic
Compass
1100's

Safety
Pin
1849

Sewing
Machine
1849

Pneumatic
Tire
1845

Safety
Match
1844

Telelgraph
1836

SafetyElevator
1853

Hypodermic
Suringe
1853

Internal
Combustion Engine
1860

Dynamiter
1867

Typewriter
1867

Barbed Wire
1873

Telephone
1876

Phonograph
1877

Polaroid Land
Camera
1947

Alomic
Reactor
1942

Jet Engine
aircraft
1939

Xerography
1937

Radar
≈ 1935

Flucrescent
Light
≈ 1935

Transistor
1947

Wonder Drugs
Antibiotics
1930-1960

Digital
Computer
1954

Wankel
Engine
1956

Artificial
Satellite
1957

Laser
1960

Polio Vaccine
1952-1955

Fig. 1-2. Important inventions in the history of mankind.

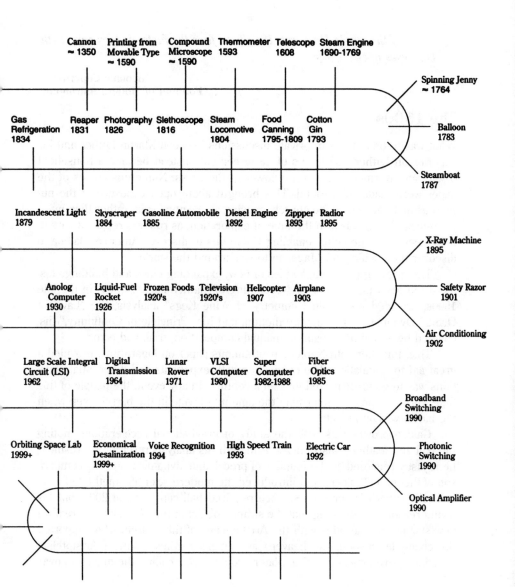

mony of the whole Universe, if only we face the facts, as they say, 'with both eyes open'."

Nicolaus Copernicus
Founder of Modern Astronomy

The 1500s

What was to be a thesis of private discussions between Martin Luther and his superior on Luther's 95 points of disagreement, instead became a household discussion overnight, due to the power of the press. Numerous copies of the paper were made by a friend. This brought about open exposure of the numerous malpractices of the church, causing an eventual split, rather than internal reform. But it was also the time of exploration, as rudders replaced oars to help steer the large ships, enabling Columbus to discover America, Balboa to discover the Pacific, and Magellan to sail around the world.

The growing new complexities of new, expanding cities and buildings fostered the need for a better understanding of mathematics. Kepler, in the late 1500s, provided his "Solid Geometry of Wine Kegs" analysis, the "Laws of Orbital Dynamics" on planetary motion, and his "Principle of Continuity," discovered by a lifetime of painful, manual computation, trial, and error.

This, together with Napier's posthumous paper on logarithms, provided a great aid to calculation. The connection with exponents and differential equations was to go unsuspected for half a century. In any event, the people of this time were now able to determine the amount of wine in the barrel, even when the barrel was on its side.

Galileo provided us with the study of gravitational accelerations, noting the ball and feather consistencies. Although his analysis resembles Oresme's, he is justly regarded as the founder of precalculus dynamics, and his comparison of theory and experiment introduced the modern scientific method of analysis. His superb lectures on science required hall capacities of 2000, but his sound reasons for thinking that the strange object in the 1604 sky was really a new star put him at odds with the Aristotelians of the church, who supported the changelessness of the heavens as being a cardinal point of Aristotelian doctrine. Astronomers 350 years later, would agree, calling the object a nova.

The 1600s

During the 1600s, the compound microscope was born, followed by the telescope in 1608, which showed that Mars had craters, Jupiter had satellites, and the Sun had spots, causing further discontent with the prevalent view of the universe.

This century became the "Age of Reason." During this time, Newton was born on a farm on Christmas day, "small enough to fit in a quart mug." He showed such interest in books that he was sent to Cambridge at an early age. However, during the Plague, the university was closed and Newton returned to work in isolation. Here, the 23-year-old developed calculus and cornered his laws of motion and gravitation, earning him fame (knighthood) and fortune. During this same period, Pascal gave us his contributions to projective geometry, rivaling his teacher Desargues. Fermat gave us the first of the great variational methods of mathematical physics "Principles of Least Time," which today remains an adequate basis for geometrical optics; but, Fermat's major, overshadowing achievement was becoming the father of modern number theory.

During this period, many mathematical principles were identified, as John Locke wrote in his *Essay Concerning Human Understanding*, "hoping to remove some of the rubbish, that lies in the way of knowledge." During this period, Bach was born, the Pilgrims landed at Plymouth, and Captain Kidd buried his treasures (1690).

The 1700s

Built upon the work of those prime mathematicians, numerous inventions occurred in the 1700s: Watt's steam engine (1767), balloons (1793), the steamboat (1787), and the cotton gin (1793). Each brought with it significant changes for society, paving the way for "the age of inventions" in the 1800s, as we entered the Industrial Revolution.

Again, mathematics was essential, as deeper theories and laws were discovered to establish the path for further advances in society. During this time, Euler created mathematics with a baby on his lap and children playing all around—pious but not dogmatic. He even withheld his own work on calculus of variations so young Lagrange could publish first, saying that "the path that I followed will be of some help, perhaps." Generations of mathematicians followed LaPlace's advice, "Read Euler, he is the master in all." Though totally blind for the last 17 years of his life, he provided an estimated 886 works, enough to fill 80 large books. From his inspiration comes the LaPlace equation, asymptotic and Mittag-Leffler expansions, Fourier coefficients, Lagrange multipliers, the Cauchy-Riemann equations, and their bearing on complex integrals, the gamma function, spherical representation of space curves, the integrability condition for orthogonal trajectories, and on, and on. And society changed, as demonstrated by the Declaration of Independence in 1776, the revolution in France in 1789, the election of Washington as president in 1789, the beheading of Marie Antoinette in 1793, and Napoleon becoming ruler of France in 1799.

The 1800s

As the newly formed hospitals provided clinical treatment of the masses—thanks to Benjamin Franklin's influence in France—the application of statistics and probabilities to medicine led to a scientific attack on cholera, as the disease centered along the water in the suburbs of London. Here, the link to sewage was discovered, causing massive changes in London's sewage system. Hence, modern medicine began as diseases were identified and classified. It became an age of separation, identification, and classification for everything from mushrooms to truffles, as botany, biology, and fossils were pursued to show not only "what was here, but what happened here." From this came vertebrate zoology and Darwin's theory of evolution.

At the same time, Gauss, the son of a bricklayer, was not yet three years old when he corrected his father's computation of a payroll. Later, he settled a 2,000-year-old question by constructing a regular polygon of 17 sides with ruler and compass, as he determined precisely how regular polygons are so constructed. He subsequently developed "the method of least squares" and proved "the law of reciprocity," which had baffled Euler and Lagrange. His *Disquisitiones Arithmeticae* of 1801 is the most important book on number theory ever written, even though Gaussian integers and their bearing on biquadratic reciprocity were still in the future.

During this period of inventing, we learned more about mechanics, thermodynamics, aerodynamics, electricity, magnetism and light from Maxwell's Equations and the inventions of Faraday (1831), Morse (1840), Edison (1869–79), Bell (1876) and Marconi (1895), as we obtained: food canning (1800), the steam locomotive (1804), the stethoscope (1816), photography (1826), telegraph (1836), the sewing machine (1846), the safety elevator (1853), the hypodermic syringe (1853), the internal combustion engine (1860), dynamite (1867), the telephone and the typewriter (1876), the phonograph (1877), incandescent light (1879), the skyscraper (1889), the gasoline engine (1885), the diesel engine (1892), radio (1895), and the X-ray machine (1895). The American industrial revolution in the North moved into full swing after the Civil War. Hence, the cities swelled as the exodus began from rural communities. This trend continued up to the turn of the millennium, with only 3 percent of the population on the farm, 75 percent in the new major cities, and many of the rest in cities having a population of 2,500 or more.

The 1900s

This was the century of change and movement, as we obtained a new breed of great mathematicians such as: Einstein, who put forth his theory of relativity; von Neumann for his theory of games, computer technology, and theory of automation; the Wright brothers for aerodynamics; and Lee DeForest and Fleming for vacuum tubes. This was the century of the nuclear bomb, the age of expanding Bell's, Edison's, Marconi's and Stroger's accomplishments to achieve automatically switched telecommunications, television, AM-FM radio, air conditioning (1902), airplanes (1903), helicopters (1907), street lights, liquid fuel (1926), frozen foods (1920s), analog computers (1930), fluorescent light (1935), cathode ray tubes (1935), radar (1935), xerography (1937), jet engine aircraft (1939), atomic reactor (1942), transistor (1947), digital computer (1950), polio vaccine (1952), artificial satellite (1957), laser (1960), integrated circuits (1962), lunar rover (1971), computer chips (1978), VLSI chips (1981), and over the recent 1960–1999 period, high-level programming languages—LISP, Fortran, COBOL, PLI, APL, C, C++, etc. (See FIG. 1-2.)

The third millennium

So it was . . . so it is . . . the light dawns and the darkness is swept away. As we ask questions, we obtain answers that change society but give us new questions. Is there any direction to our route to knowledge or do we make up the route as we go along? As we achieve one more step to understanding, our knowledge changes. As our knowledge changes, so do we. . . .

As the world population expands, we're faced with new challenges, problems and opportunities. As our personal understanding and relationships expand to include our human, environmental, global and universal relationships, our understanding and appreciation of the possibilities that technology offers will also increase and expand. (See TABLES 1-1 through 1-2B.)

In conclusion

Now that we've paused to review our progress, the role of technology becomes quite evident. We've seen how each society's technology has provided the basis for its following society, whose technology provides the change and support for its subsequent society. In reviewing this cycle, we can't help but ask the inevitable: Where do we go from here?

In considering the history of inventions, shown in FIG. 1-2, it's interesting to note how technology has come in waves. First the mathematical discoveries

Table 1-1. Population Growth

Location	Population–1975	Pop./Sq. mi.	Growth	Population–2000 (est.)
World	4,105,000,000	—	2.1%	7,000,000,000
Africa	423,000,000	$(36/m^2)$	2.8%	—
Asia	2,395,000,000	$(142/m^2)$	2.3%	—
Australia	14,000,000	$(5/m^2)$	1.6%	—
Europe	676,000,000	$(166/m^2)$	.7%	—
North America	351,000,000	$(36/m^2)$	1.3%	—
U.S.A.	218,364,000	—	2.2%	280,000,000
South America	236,000,000	—	2.9%	—
China	855,000,000	—	—	—
India	624,000,000	—	—	—
Russia	259,000,000	—	—	—

Table 1-2A. Shifting Challenges

- Globalization versus localization
- Infolopolis versus megalopolis
- Rural versus urban cities
- Planned versus unplanned cities
- Balanced versus unbalanced geographical populations
- Balanced versus overloaded ecological systems
- Clean versus unclean environment
- Synergistic utilization versus underutilized ecosystems
- Sectionalized versus nationalized regions
- America versus Europe versus Asia versus Africa
- Global versus national government
- Electric versus gas/gasoline transport
- Electronic versus physical transport
- Controlled versus random genetics
- Solar/nuclear versus fossil energy
- Religious-versus nonreligious-based society
- Disciplined versus undisciplined society
- Moral versus amoral society
- Family versus nonfamily community
- Group versus individual
- Personalization versus depersonalization
- Reality versus simulation

Table 1-2B. Opportunities

- Genetic food growth and control
- Undersea harvests
- Sea water desalinization
- Weather control
- Global greenhouse effects control
- Solar energy
- Nuclear energy
- Electrical energy
- Supersonic transport
- Bullet trains
- Electronic medical diagnostics
- Disease elimination
- Global communication
- Colonization of local planets
- Deep space exploration

provided the fundamental theorems. Then, the engineers and inventors applied these scientific possibilities to market opportunities. This then fostered a new growth in society by meeting the changing needs of an expanding population.

Today, we're at the beginning of a new wave. We'll see, in the next chapter, how two great technologies are now ready to launch us into a new era of products and services that will meet an exploding population growth. The world population will begin doubling every twenty or so years. This doubling of billions, not millions, will require new mathematical advances, new technologies, new inventions, to meet new billions of human needs.

Hence, the next millennium will be "The Globalization Age," as we use these possibilities to communicate and interrelate from our homes and businesses to anyone, anywhere, anytime to achieve the The Information Age—The Information Era.

"With the splitting of the atom everything is changed, except our thinking . . . unless we think anew, we face catastrophe, and that's our challenge . . ."

Albert Einstein

2

Information technologies

"There is a new frontier,
A new ocean to sail;
Let us launch a ship to sail it . . ."

John F. Kennedy

The years of mathematical achievements have now set the stage for the third millennium. With the emergence of the computer and communications technologies, each in its own right considerably sophisticated, we're now at the moment of their integration and widespread use in every facet of society.

So, let's see how far technology has progressed. What does it offer, and how can we use it successfully?

It's hard to believe that it was only in the mid '50s that freshman students in engineering laboratories of the more advanced universities were learning how to program the analog computer. They used machine-coded numbers to formulate shift left, shift right, and adder instructions to solve the mathematical equations resulting from the work of Euler, LaPlace, and Fourier, as they studied Bernoulli's flow mechanisms, Newton's laws and Fermi's optics. It was not until their junior and senior years in the late '50s that future electrical engineer's oscilloscopes moved from looking at signals passing through De-Forest's diode and Fleming's triode tubes to analyzing the new transistorized modulation and demodulation techniques for communication and signalling. This then led to the newly developed transistor/trigistor-type on/off decision-making properties of wired logic, as emphasis shifted to digital computation, command, control, and communication systems for military applications.

The military led the way in digital technology in the early '60s, with Minuteman designs, using a newly developed device by Texas Instruments called an integrated circuit. It usually provided two major or four minor decision-making registers or logic gates, at a cost of $55 per chip in lots of 1 million. As

West Coast military designers pursued their accomplishments using logic tables, East Coast designers in the commercial arena used logical flowchart mechanics to define their Boolean equations for logical decisions, as they developed such systems as the IBM 360 computer with its FORTRAN, APL, RPG, and COBOL higher-language software.

Hence, the '60s were the years of early large-system hardware and software developments, as advances in key computer and communication technologies were achieved in not only the IBM 360 computer and Minuteman parallel processing computers, but also in the #1 and #2ESS, Autovon, Autodin, 465-L, and 490-L, SAGE, Safeguard, and Appollo voice and data networks. These enabled the American president and his generals to talk to their Soviet counterparts, as well as their SAC command bases. So passed the '60s. As Cold War tensions mounted during the Cuban crisis, test missiles at Vandenburg and other Air Force bases were reprogrammed with new "test sites." Then, as Robert McNamara said to the aerospace-military industries, "though I have built you, I will not sustain you," so began the long process of technology transfer from West Coast design centers in Los Angeles and Seattle to Apple in Silicon Valley, Bell Labs in Naperville and New Jersey, Control Data in Minneapolis, Digital in Boston, IBM in New York and San Jose, GTE in Boston and Chicago, ITT Technology in Connecticut, and on to various sections of the world as the military/aerospace LSI and VLSI designers moved here and there to use their technologies and capabilities in commercial endeavors.

Computers and communications

By looking at the impact of past technology and showing the resulting affect on our way of life, we can begin to appreciate technology's future impact on society. This has been the purpose of the preceding analysis. So, let's take a more detailed look at the evolution and resulting capabilities of computers and communications to more fully understand not only why, but where and when we could/would/should deploy these technologies from this time on . . .

During the '60s, antitrust concerns between IBM and AT&T drastically impacted the direction of communications. As a direct result of these concerns, as well as having a limited knowledge of future computer needs, AT&T elected to ignore the data world and concentrate on voice communications. As an afterthought, and usually only as a result of seeing many data transport customers lined up outside their doors demanding some form of data handling, AT&T, GTE and the other common carriers provided a family of very low-speed data modems, "data sets," for moving 600, 1200, 2400, and 4800 bits per second. (Remember it takes seven or eight bits to represent a character such as the letter A.) This was a small, evolutionary advancement from the telex rates of 60 or 100 words per minute (see Appendix—Data Users).

Hence, users were left to only a little more than ticker-tape rates, as they patiently (or impatiently) waited for the faster dissemination of information. Costs were substantially higher for higher bandwidths, since transport was based upon considering the higher bandwidth throughput in terms of its equivalent number of transported voice calls, and not by the service it provided. Hence, the picturephone offering of the late '60s and early '70s was quite limited in resolution and quite expensive to use, especially because as the distance between users grew and grew, so did the costs of service.

During the '70s and '80s, computer capabilities, measured by some in terms of instructions processed per second and by the size of internal memories, grew in leaps and bounds, as these measurements initially doubled every two to four years and then every 6 to 18 months. In this manner, mainframes shifted from batch processing (where trays of cards were rolled into the computer room) to front-end, stand-alone systems, and then to remote, interactive distributed systems.

As transistor technology shifted to LSI (large scale array integrated circuits) to VLSI (very large scale integrated circuits), so the number of logic decision-making gates and storage devices, contained in an area the size of the tip of a pen, increased and increased again to the point of hundreds of thousands of decision elements. Here, customized boards became customized chips, only to change again to the "computer in a chip—memory in a chip" technology, thereby enabling millions of instructions per second and millions of storage areas for processed information. This technological opportunity created a pressure bulge that could not go away (as even a tightly contained protruding stomach eventually pops out here or there). So it was with the pressures created to apply this advancing technology.

As these computer chips became more and more economically available, they could be purchased by the general public and industry and applied to every application in the marketplace. Pressure then increased for a computer in every office and one in every home, as did the pressure to interconnect them. Internal problems were initially resolved by extending the high-speed bus technology in the computer to its local users. This became the local area networks promoted as Ethernet by Digital or SNA by IBM.

To interconnect clusters of users on these 1 to 2 million bit-per-second local buses (later to evolve to 10 million bits per second), bridges were developed to cross over to adjacent local area networks. As layer addressing became more complex, routers were developed to route the calls to several local networks within the complex, across town, or even access the country, using leased line communication facilities to interconnect these point-to-point transports.

In the '70s, data switches were developed to enable point-to-multipoint, as well as any point-to-any-point dynamic switching, using switched connections on a dedicated circuit. Or, information could be independently moved on a usage basis over shared, less expensive media in the form of packets, using store and forward algorithms. Packet sizes increased from 256 characters to 1,000 to 64,000 character sizes, as pieces of messages were interlaced with other messages. Here, users only paid for the amount sent, not for holding a path to the destination for the full period that they might wish to communicate.

Hence, only when they talked or actually sent information was the path held, not during the idle or think time.

During the '80s and into the '90s these dedicated circuit and usage packet-handling techniques were being perfected, as transport throughput rates increased and increased again. So much so, that new technologies to interconnect local area networks expanded from 1, 4, 10, and 16 million bits per second to 100 and 200 million bits per second. This then led to the entrance of directly switched broadband paths between users at the 50, 155, and 600 million bits per second rates.

However, the marketplace continued to note the need for the slower but more functional, versatile and ubiquitous data-handling rates of 64,000 and 128,000 bits per second for the majority of users, as the masses collected and distributed information through inquiry/response mechanisms to determine the availability of theater tickets, credit limit, bank balance, whereabouts of a piece of merchandise, etc.

Hence, there came a need for narrowband transport containers for many of the data users: wideband containers for those on internal business LANs (where they're coalesced in groups), broadband containers for the movement of graphic images, continuous video movement, and high-speed bulk, computer-to-computer information exchange in which large databases are searched, shared, and expanded.

So by the mid-'90s, interdependent computer information access, search, storage, manipulation, and presentation will have become more and more dependent on three somewhat distinct but evolutionary narrowband, wideband, and broadband transport mechanisms, as the integration of C&C (computers and communications) becomes complete in the form of P&P (private and public) networking and internetworking.

Narrowband—wideband—broadband

Narrowband data has not advanced in America as quickly as it has in Europe, where, by 1990, each Postal Telephone and Telegraph (PTT) had its own public data network. America has, for regulatory and antitrust reasons, mainly concentrated on providing voice-grade services. This path has led many a good network marketer to say, "Why not just let the data users simply use their voice grade modems? Why should we provide them a separate low-speed data network? When enough data users request ISDN, and if and when it's available, we'll use it!" (*ISDN* is the integrated voice and data network that was delayed for years because it was mainly sold as an interface, not as a new data network being added to a new premium voice network). This wait and see, don't make waves, don't bother me with some excitingly new and different approach, has unfortunately delayed the introduction of the lower-speed public data network, called narrowband Integrated Services Digital Network (n-ISDN).

All of this has continued the push and drive of potential data customers from the public network to the private.

In the past, the traditional public network providers mainly responded to voice-handling demands with voice service variations of existing services, such as: Touch-Tone instead of dial pulse, or voice mail instead of terminal recorders. Initial attempts to simply sell ISDN as a second voice line, with a future service statement indicating the possibility of new data-handling capabilities, didn't generate an immediate, overflowing demand for ISDN. Many of its potential data offerings were not deployed, as many RBOCs (regional Bell operating companies) sat on the fence to assume a wait and see attitude.

Competitors recognized the opportunity to bypass public networks with private data networking nodes, and that awakened new interest in data in the early '90s. New offerings initially centered only on wideband point-to-point data transfer to encourage more economical bulk usage transport, rather than using the expensive dedicated circuits. This was to compete head to head with private users, who had discovered how to economically use 1.5-million-bits-per-second T1 facilities or even 45-million-bits-per-second T3 facilities to switch their various local and remote campuses of users here to there, using more or less capacity as application needs and destinations dictate.

Hence, the '90s formed the battle lines between private and public networks, as well as between RBOCs and VANs (value added networks) provided by enterprising alternative providers (APs), as customers desired to access more and more remote database services (DBSs). So, new transport protocols for interconnecting local area networks were deployed for wide area networks (WANs) within a local campus or local community. These protocols include: Frame Relay, a variable-length packet transport mechanism, SMDS, a metropolitan area network fixed-cell packetize transporter, or FDDI, a private 1 to 100-million-bit private internetworking transport protocol.

This rush of activity and opportunity led network and marketing providers to observe that transitional grouping of bandwidth could be provided to the users from internal rates of 64/128K bits per second to multiples of 64K to the T1 or T3 rates, and on to broadband higher rates of 50+ Mb/s. Hence, the full data-handling switched transport range reaches down to include the basic 64K/128K low-speed user. By conditioning the current network (plant), it can deliver significantly higher speeds of information—as we expand rates. Therefore, in going from 64K/128K to multiples of 64K up to T1/T2 (6.4M b/s), this is the beginning of a series of evolutionary moves, still using existing plant, but advancing its capabilities. Selectively deploying T3 (45 million bits per second) is another jump requiring coaxial or fiber-optic capabilities. However, fortunately or unfortunately, depending on one's perspective, nothing is ever quite that easy. We need to use what we have for today, but we must also begin deploying what we'll need for tomorrow. That will require revolutionary changes. Eventually, the preferred vehicle for the public network will be a single-mode fiber, which can support numerous frequencies of varying wavelengths to initially achieve the capability of 6,000, then 18,000 (three frequencies), and so on to 100 or more varying wavelengths. Hence, the capacity of a single fiber appears unlimited. However, it will be a long time (2015+) before there's a fiber to every home.

Actually, today's ubiquitous plant will adequately support narrowband traffic from the low-to-medium-speed data users. (See appendix—The global information marketplace—today and tomorrow.) Today's plant will make a nice, realistically inexpensive base for transporting low-speed 64K/128K b/s traffic (which is significantly enhanced from previous data rates of 600, 1200, 2,400, 9,600, 14.2K, 19.1K b/s). What's needed is the additional numbering, addressing, error correction/detection, alternative routing transport capabilities, and the additional value-added capabilities of delayed delivery, broadcast, and polling. This will result in a universal public data narrowband offering that will be substantially more beneficial to the general data user, especially if priced appropriately.

Alternatively, the broadband network of the future will meet the needs of video users, especially as large mainframe supercomputers, such as "the Cray," process information at 200+ million instructions per second (MIPS) and interoperate with sophisticated 20+ MIPS workstations or other supercomputers to provide extensive graphics and computer analyses for weather maps or for building automobiles, bridges, and skyscrapers. Similarly, X-ray images can be transported over switched broadband facilities to better enable remote diagnostics by specialists, located at the Mayo Clinic in Rochester, Minnesota, or Johns Hopkins University in Baltimore, Maryland, or St. Vincent's in Dublin, or . . .

Finally, videophone should be available in all three entities, as its resolution, gray scale, color, motion, response, and overall quality improve as its images are transported over narrowband, wideband, or broadband facilities. Similarly, video conference centers can operate over wideband or fully broadband transports, in order to provide an alternative to long-distance travel.

Figure 2-1 indicates these three major functional areas—narrowband, wideband, and broadband—in terms of public voice, public data, public wideband, and public broadband network's features and services that will satisfy various inquiry/response, data collection, data distribution, and processing application's needs. We'll see later how the various industries will usually require new transport capabilities, features, and address services from several of these networks for their specific applications.

These networks can be established using public facilities. Here, new switching nodes can exist on customer premises, as new broadband switches replace traditional PBXs (as fifth-generation physically distributed systems). These CPE systems will interface to new network access nodes for distribution and transfer. The local network will consist of survivable and secure rings, homing on multiple, higher-level systems (in the event of failure) and long-distance interexchange carriers points of presence. New superswitches will be

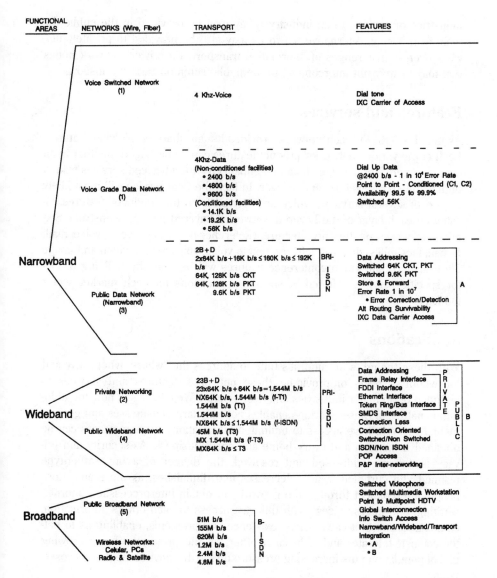

FUNCTIONAL AREAS	NETWORKS (Wire, Fiber)	TRANSPORT	FEATURES
	Voice Switched Network (1)	4 Khz-Voice	Dial tone IXC Carrier of Access
	Voice Grade Data Network (1)	4Khz-Data (Non-conditioned facilities) • 2400 b/s • 4800 b/s • 9600 b/s (Conditioned facilities) • 14.1K b/s • 19.2K b/s • 56K b/s	Dial Up Data @2400 b/s - 1 in 10⁴ Error Rate Point to Point - Conditioned (C1, C2) Availability 99.5 to 99.9% Switched 56K
Narrowband	Public Data Network (Narrowband) (3)	2B+D 2x64K b/s+16K b/s ≤ 160K b/s ≤ 192K b/s [BRI-ISDN] 64K, 128K b/s CKT 64K, 128K b/s PKT 9.6K b/s PKT	Data Addressing Switched 64K CKT, PKT Switched 9.6K PKT Store & Forward Error Rate 1 in 10⁷ • Error Correction/Detection Alt Routing Survivability IXC Data Carrier Access [A]
Wideband	Private Networking (2) Public Wideband Network (4)	23B+D 23x64K b/s+64K b/s=1.544M b/s NX64K b/s, 1.544M b/s (f-T1) 1.544M b/s (T1) 1.544M b/s NX64K b/s ≤ 1.544M b/s (f-ISDN) 45M b/s (T3) MX 1.544M b/s (f-T3) MX64K b/s ≤ T3 [PRI-ISDN]	Data Addressing [PRIVATE] Frame Relay Interface FDDI Interface Ethernet Interface Token Ring/Bus Interface [PUBLIC] SMDS Interface Connection Less Connection Oriented Switched/Non Switched ISDN/Non ISDN POP Access P&P Inter-networking [B]
Broadband	Public Broadband Network (5) Wireless Networks: Celular, PCs Radio & Satellite	51M b/s 155M b/s 620M b/s 1.2M b/s 2.4M b/s 4.8M b/s [B-ISDN]	Switched Videophone Switched Multimedia Workstation Point to Multipoint HDTV Global Interconnection Info Switch Access Narrowband/Wideband/Transport Integration • A • B

Fig. 2-1. Narrowband, wideband, broadband.

deployed for external routing, address translations, and network control. (See appendix—Private and public internetworking.)

On top of these transport layers will sit information platforms, generally referred to as service nodes or information switches, to expand traditional Centrex/Centron type offerings for closed user groups and provide gateway services to Application Service Center's databases—which may be common across industries or specific to an industry. They may be provided in the public domain or privately. Alternative networks may also be provided to privately network over similar ranges of information transport, via private network nodes that may or may not interconnect to the public common carrier transport.

Features and services

As noted in FIG. 2-1, each network provides its fair share of evolving features. Each of the various networks' product features can be packaged together in an overall generic service, which can be enhanced with advanced services from a higher transport level or service layer. In fact, the total service will most likely consist of several services, which are derived from the products' features, located on each layer of the Layered Networks' Layered Services construct. See previous works on this architecture (References). In any event, voice mail, E-mail, facsimiles, delayed delivery, text-to-voice, voice-to-text, video and image file transport, and video conference features will become the building blocks for specific services, offered to specific applications for each market sector industry.

Applications

With these thoughts in mind, it's time to address the "where, when, why and for whom" aspects of providing these networks' features and services to eighteen or so major industries and user groups. We'll later see how the ubiquitous nature of these offerings enables not only large businesses and governmental users to have access to unlimited information, but how access can be obtained from each and every home and small business. As security and privacy issues are addressed and resolved, the danger of a big-brother-type planned and curtailed society decreases and diminishes, as more and more people interconnect through the networks to obtain interprocessing, internetworking, and interservices. With this vast access to each others' information, comes awareness of each others' existence and problems, enabling us to better coexist, tolerate, and help each other, as the presence of our growing global population puts increasing pressure on Earth's environment and ecosystems.

This is the promise of the Information Network's technology. It's to help us meet the challenges of the next millennium, the third millennium, the "Information Millennium," with its "Global Information Society."

> Most people are willing to do great things,
> which they put off till tomorrow,
> but few are willing . . .
> . . . to begin doing them today . . .

Part II

Information applications

Information usage
begets
Information usage
begets
Information usage

3

Information applications' services

Without a need, there is no will;
With a need, there is a will;
The greater the need, the greater the will.
Where there is a will, there is a way.

As suppliers of new communication systems become more and more focused on the needs of service providers, it becomes increasingly important for suppliers to accurately represent both the service providers' requirements and the needs of the providers' customers, the users of the services. But what are the needs of the providers and the providers' customers? For many years, traditional communication common carriers have focused on voice services. We all know what voice dial tone provides, but do we know what data or video dial tone provide? So, where do we begin? Let's begin with needs. Willingness to pay is directly proportional to the size of the need. We've all seen the consumer willingness to spend that has resulted in the rising costs of housing. But we've also seen falling markets due to improper house design, wrong location, or more personal issues such as fear of layoffs, recession, and concern for the future. To be successful during difficult times, builders must design the right houses for the right customers in the right location. Thus, timely solutions must be relevant to realistic applications. The solutions must address and resolve changing customer needs. (See FIG. 3-1.)

Customer needs

As noted in my previous book, *Global Telecommunications*, customers of the '90s and beyond will need to see more and more information. They'll need to

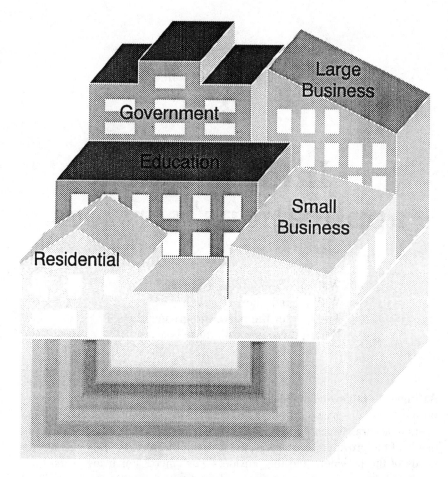

Fig. 3-1. Information applications.

see information in text form, in image form, in graphic form, and in video-phone. They'll need to manage and control the creation, storage, access, retrieval, transmittal, manipulation, processing, and presentation of information. As daily operations become more and more dependent on having the correct information on a timely basis, from any location in any form, it's necessary to foster the rapid exchange of accurately transmitted information. To do so requires diverse systems to be interconnected in a manner that ensures that information can be transported between users without undue delay. As time passes, it will become more and more important to both simultaneously hear and see information at the desired level of quality and resolution, using dynamically adaptable facilities. To achieve this, basic information transport rates must become more and more economical to encourage more and more usage.

As noted in FIG. 3-2A, customer needs for specific data-handling and processing capabilities vary from the timely arrival of information to the elimination of queues in one's daily tasks. No one really wants to drive to the bank and stand in line (a queue) to determine one's checking account balance. In time, we need to enable numerous inquiry/response interactions to be performed electronically from one's home or business, so remote access can be established to local databases. As one professor in a database management class of the '60s perceptively noted, "Lists will rule the world." So it may be in the late '90s.

The global information marketplace today and tomorrow

First, for the sake of clarity, let's define what's meant by the various data-handling terminologies (see appendix—Global data users). Next, let's take a closer look at the marketplace, from market needs to applications, to better understand what services are needed for the turn of the century.

Customer Needs

Timely Information	Data Transfer	Data Manipulation
Time Management	• Point to Point	Data Presentation
Queue Elimination	• Point to Multipoint	Delayed Delivery
Inquiry/Response	Data Storage	Image Generation
Data Collection	Data Access	Image Storage
Data Distribution	Data Retrieval	Video Display
		Video Entertainment
		Video Education

Fig. 3-2A. Customer needs.

Market needs

In reviewing customers' needs and current and future network offerings, it's quite apparent that the traditional "voice only" network structure must change (see FIG. 3-2B). There's a need for a continuing array of new offerings. Therefore, to support the various market sectors, in the '90s, we need:

- A structure from which to offer new services—both the regulated and nonregulated—in a timely fashion.
- A structure from which to offer enhanced voice services.
- A structure from which to offer data, image, and video services.
- A structure to enable interoperability of computers of different capabilities.
- A structure that ensures secure and survivable (S&S) transport of information.
- The internetworking of private and public networks to encourage usage and growth of shared public facilities.

In considering how we currently handle the growing numbers of data users, we see them growing from only 3 to 6 percent of the network terminals in the '70s to an anticipated 10 to 15 percent by the turn of the century. Though some might not see this as a significant amount, designers realize that the voice network can't sustain its designated voice traffic loads when handling this increased volume of data traffic over its dial-up facilities. Some estimates are that by the year 2000, the data traffic generated by these terminals will be equivalent to 50 percent of the voice. With this comes new, more complex network switching and transport requirements. During growth, users will experience many problems from both inadequate terminals and the traditional

Fig. 3-2B. Market needs.

network, causing pressure to bypass the network to private networks that more properly handle the more sophisticated terminals. These conflicts and complexities are adequately expressed in the following perspective from a frustrated and disgruntled receptionist who, besides answering telephones, was asked to handle the growing number of business faxes.

A user's perspective

"In reviewing the daily fax operation on the voice network, it was determined that thirty or more faxes were sent and received daily from the reception area. These faxes were anywhere from a single page to over fifty pages in length. With present voice grade dial-up equipment, faxes transmit very slowly. When there's a lot of writing or dark areas, it can take three minutes or more for one page to transmit.

False reports were received on the 'transmittal report' sheets that were sent after sending a fax. At times, the report would read 'good,' but later calls were received indicating that the fax was not readable and must be sent again. On other occasions, the 'report' read 'communication error,' and we later learned that the fax went through properly. These false reports resulted in unnecessary phone calls and redundant sending of faxes. There were times when a fax had to be sent several times (4 or 5 or more). This happened when faxes were not readable or when pages ran together, etc. This applied to both outgoing and incoming faxes.

The pages were numbered on the back of outgoing faxes, because they didn't come off of the machine properly. This procedure became necessary in order to return them back in the correct order to the originator. Besides transmission errors, frequent jamming of paper was a problem. Copies often had to be made before sending a fax, because if any other paper was used, the machine either rejected it or tore the edges. Jams often occurred right in the middle of faxing a document. There were times when a fax had to be hand fed one page at a time before it would transmit without jamming. Many times people just gave up and sent the material 'airborne' after trying to fax it several times, thereby incurring another $6.00 in expenses.

All of these problems made faxing documents very difficult and time consuming, as well as very costly because of the aforementioned redundancy involved."

Needs—applications—services

This bulging, blossoming, blooming, but somewhat bungling, data market provides an expanding opportunity for new types of terminals, such as dataphones, imagephones, image systems, monitors, sensors, workstations, feedback control systems, videophones, and high-definition television. With these devices available, it's interesting to consider how each major industry would use the new technology in their daily operations. Figure 3-3A notes some of these tasks. For example, automobile parts stores rely heavily on a central warehouse and fast cars to move inventory to points of sale that are distributed throughout the metro. Here, the right data-handling services can make or break this operation, as they change from voice call ups that manually read long lists of inventory numbers, to slow-speed, fax-type transport, to dial-up text order form-type mechanisms, to higher-speed fully digital "error free" data transport.

As the application tasks become more information intensive, it may be desirable to offer higher- and higher-speed solutions by moving from narrowband facilities to wideband to broadband. However, each has its own place and each its own time. It's interesting to observe, in reviewing the figures, that the applications that use more and more visual information will require more and more bandwidth; but the higher bandwidth is only needed to satisfy a service, which could not be accomplished in the same manner at lower rates. Hence, business video conference centers require broadband transport services, while point of sale terminals are nicely handled by narrowband offerings. Similarly, remote workstations will require the capability to rapidly exchange large amounts of data with more sophisticated mainframes, over wideband facilities, once their front-end processing is accomplished.

Video compression techniques can be deployed to lower bandwidth usage, but it's also true that, in many cases, this also lowers response to rapid movement and provides less overall quality resolution. In fact, as videophone codices are deployed to reduce videophone usage of T3 (45M bits per second) to 384K bits per second, not only is picture quality reduced, but, in some instances, voice no longer correctly tracks mouth movement, due to transport delays. (Of course, better algorithms and better transmission mechanisms are continuously being developed, but many of these reduction techniques simply change transmission challenges into expensive and complex CPE processing challenges.)

We've also seen how success spoils success, where a network that responds properly to a few users will "break down" once overloaded with many users. Similarly, LANs/WANs/MANs designed for one type of bursty variable-bit-rate data traffic will become less effective when challenged to meet the needs of continuous-bit-rate video users. Many times, each application will require different transport capabilities, depending upon how many messages, and what type of messages, are deployed in each particular location. Thus, as different communities of interest are addressed, different functional service solutions are needed. (See FIGS. 3-3A and 3-3B.)

Narrowband/wideband/broadband services

As FIG. 3-3B shows, each functional area could offer one or more network solutions that support a specific grouping of service offerings. In this, Videophone I is provided at the narrowband rate of 64K b/s, Videophone II at 1.5M b/s and Videophone III at 50 to 155M b/s.

Closer scrutiny shows that the basic narrowband Public Data Network

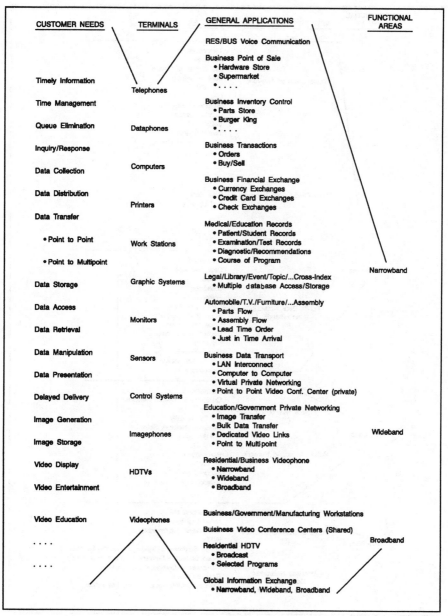

CUSTOMER NEEDS	TERMINALS	GENERAL APPLICATIONS	FUNCTIONAL AREAS

RES/BUS Voice Communication

Business Point of Sale
• Hardware Store
• Supermarket
• . . . .

Timely Information

Telephones

Time Management

Queue Elimination

Business Inventory Control
• Parts Store
• Burger King
• . . . .

Dataphones

Inquiry/Response

Business Transactions
• Orders
• Buy/Sell

Data Collection

Computers

Data Distribution

Business Financial Exchange
• Currency Exchanges
• Credit Card Exchanges
• Check Exchanges

Printers

Data Transfer

Medical/Education Records
• Patient/Student Records
• Examination/Test Records
• Diagnostic/Recommendations
• Course of Program

• Point to Point

Work Stations

• Point to Multipoint

Narrowband

Data Storage

Graphic Systems

Legal/Library/Event/Topic/...Cross-Index
• Multiple database Access/Storage

Data Access

Automobile/T.V./Furniture/...Assembly
• Parts Flow
• Assembly Flow
• Lead Time Order
• Just in Time Arrival

Monitors

Data Retrieval

Data Manipulation

Sensors

Business Data Transport
• LAN Interconnect
• Computer to Computer
• Virtual Private Networking
• Point to Point Video Conf. Center (private)

Data Presentation

Delayed Delivery

Control Systems

Education/Government Private Networking
• Image Transfer
• Bulk Data Transfer
• Dedicated Video Links
• Point to Multipoint

Image Generation

Imagephones

Wideband

Image Storage

Residential/Business Videophone
• Narrowband
• Wideband
• Broadband

Video Display

HDTVs

Video Entertainment

Business/Government/Manufacturing Workstations

Video Education

Videophones

Business Video Conference Centers (Shared)

. . . .

Residential HDTV
• Broadcast
• Selected Programs

Broadband

. . . .

Global Information Exchange
• Narrowband, Wideband, Broadband

Fig. 3-3A. Needs and applications.

provides a whole host of data transport services that support the "data handling" of text information. With this network, messages are broadcast, blocked, delayed delivered, routed over alternate routes, secured, or converted to enable access to different systems or specialized databases.

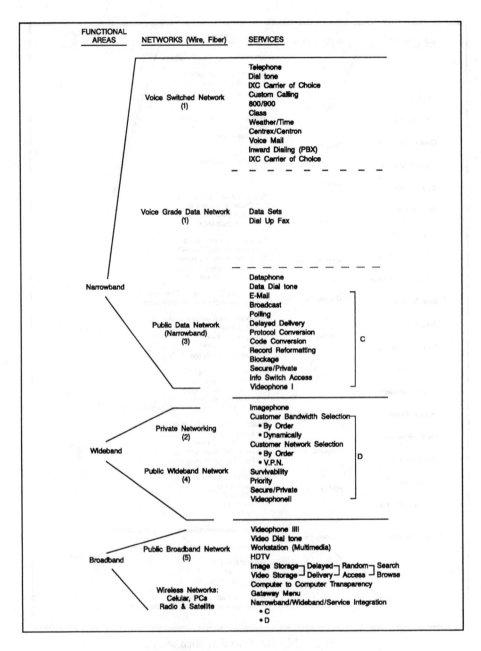

Fig. 3-3B. Functions and services.

To further appreciate the vast array of new services, appendix A provides a more detailed look at narrowband, wideband, and broadband services. The tentative lists shown in FIG. 3-3B are expanded to encompass the full spectrum (FIG. 3-4) of opportunities spanning a wide range of technical possibilities.

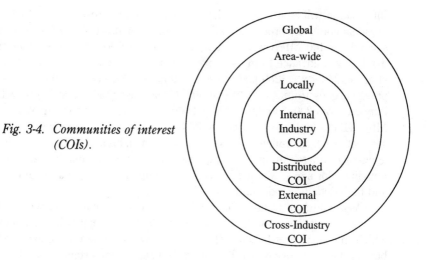

Fig. 3-4. Communities of interest (COIs).

Information services applications

As the communications industry turns to the information marketplace to best determine which technology should be applied, it's essential to fully understand and appreciate all aspects of the various applications in order to identify what's applicable today and tomorrow. The term *application* can be defined as an act of putting to use (of new techniques) or a use to which something is put (new for old remedies). Hence, to decide what's applicable—capable of or suitable for being applied—we need to look at each major market sector and consider its essential tasks, its mode of operation, and its community of interest in terms of its movement and manipulation of its information.

Information is "handled" in many ways, as it's created, captured, accessed, stored, retrieved, searched, screened, browsed, indexed, compressed, organized, published, presented, manipulated, computed, communicated, cross-related, sequenced, related, documented, associated, visualized, interrelated, walked through, blocked, windowed, randomly located, and presented.

These many different ways of handling information enable computerized visualization, full context indexing, nonsequential listing, photorealistic graphics, hypertext, hypermedia, multimedia, 3-D imaging, relational access, electronic imaging, computer imaging, virtual reality, virtual imaging, and various forms of information storage and retrieval from laser discs, video discs, compact disk-read only memories (CD-ROM), CD-I (interactive), dynamic random access memories (DRAM), and write once read many times (WORM) memories.

This will achieve full motion real time video (RTV), production level video (PLV), solid modeling programs, geographical information system databases (GIS), computer art, animation graphics, data storage, and imaging systems.

Information will be available in low-resolution to high-resolution color form, as personal computer processing advances from 2 MIPS (million instructions per second) to 100 MIPS workstations to 1000 MIPS knowledge stations. Here, displays will advance from 640 <x> 480 pixels to 1280 <x> 1024 pixels, then shift to 1280 <x> 1280 pixels, then change to 4000 <x> 4000 pixels as compression goes from 2:1 to 10:1 to 30:1 to, perhaps, 100:1. Hard copy will also improve from 200–300 dpi to 600 to 2000 dpi, as fiber transport moves from 90 Mb/s (1980) to 2.4 Gb/s (1990) to 20 Gb/s (2000). This will enable many new services, such as: bulletin boards, electronic mail, online conferences, file management/storage, foreign language translators, medical imaging, architectural imaging, plot maps, weather maps, solid mapping, high-speed documentation, and high-definition videophone.

As we attempt to take a closer look at what information services will satisfy each task, it's important to recognize that not only will the manner in which information is handled change, as new technology becomes available, but the tasks themselves will change. Hence, it's important to not only consider information in the form of basic data, but also as text, image, graphics, and video. Its use and deployment will be interactive or broadcast, continuous or bursty, as a singular independent inquiry or a series of dependent inquiry/ responses—as high quality, high resolution, or as having a tolerance for transport errors—using error detection/correction methods. It will be transmitted as secure and protected information or somewhat open or unprotected, as highly private and unavailable, or available, as sensitive or insensitive to various types of message transport access or delays. In this manner, the best type of communication can be determined for each task, or at least the full range of opportunity is so noted, as we consider both the evolutionary and revolutionary aspects of deploying narrowband, wideband, and broadband technological solutions over the next twenty years.

Therefore, as we look at technology, we need to identify types of usage by technical characteristics and attributes, which differentiate what will be available, where, and when. Here, selected or ubiquitous, expensive or inexpensive, today or tomorrow . . . deployment considerations play a major role in establishing appropriate pricing strategies to ensure that either an offering has external growth and full use or is tightly controlled and is only available for specialized, selected, closed user groups.

Community of interest

Each industry's community of interests will need to be identified internally and externally within both local and global arenas (see FIG. 3-4). Then we'll see a spiraling array of involvement, as users within a local complex interface to external users within the local community or across a region, nation, or the globe. For example, we need to determine how medical doctors within a hospital exchange information among themselves and other doctors throughout the community, as well as specialists across the nation. In this manner, we can better understand how communication can help them examine X-rays, access patient files, or obtain information from distant databases, such as those that identify the side effects of particular drugs or the use of particular practices and procedures.

It's with these considerations in mind that we'll briefly review the major applications in order to help determine what might be needed and when. In so doing, each community of interest will be explored to show the full scope of information exchange to better appreciate, understand, and qualify the magnitude of the opportunities as well as the phases of change. The quantification of the demand for subsequent analyses remains with future market researchers to pursue cross-elasticity of demand, macro-micro service models, or the plain old "give it to them, let them try it and see if they like it" marketing analysis techniques. Finally, after contemplating the assessment of each market, we need to step back and observe commonality and interdependence across market sectors, industries, groups and users. This will help indicate the need and urgency for developing new capabilities selectively or ubiquitously on a specific or shared basis.

The purpose of the following specific industry application analyses is to enable suppliers, providers, users, and regulators to better appreciate the present and forthcoming technical possibilities and market opportunities of these specific applications.

Information industries

As we turn to applications, it's now time to study the possibilities and opportunities of the following various industries—banking, investment, wholesale/retail, insurance, law agencies, city/state government, federal government, manufacturing, utilities, transportation, health care, education, news/magazine information exchange, entertainment, small business, home communications, and general information services—in terms of various data-handling modes of operation, such as: inquiry/response, data collection, data distribution, interactive remote access/time sharing, remote documentation/display, interactive graphics, transactions, as well as image and video presentation.

Information applications model

As FIG. 3-5 indicates, each industry can be assessed from several perspectives, and then reassessed in terms of its cross-industry commonality. In this manner, the individual community of interest network for each industry can be discussed in terms of the tasks that satisfy its business operations, the communications and computer services that assist in completing these tasks, and the narrowband, wideband, and broadband networking needs of each new group of the industry's information users.[1]

---

[1] This model will be further clarified as we consider the following applications.

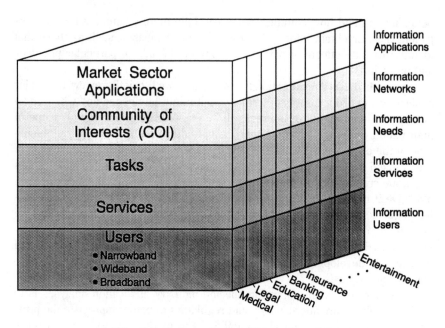

Fig. 3-5. Information applications model.

Once identified for all industries, this new vantage point will enable interesting reassessments of these entities in terms of common and not-so-common network configurations, information-handling, information services, and information users' perspectives. So now let's proceed with a look at each major industry to determine how it will use and apply the forthcoming information infrastructure.

Market sector—industry—product

As we consider the information marketplace, it can be viewed in terms of general market functional areas that are further separated into industries and their products. This classification and differentiation is seen as we proceed deeper and deeper into our understanding of customer's needs, as we attempt to identify their mode of operations, their daily tasks, their type of communication - information handling that they need to complete their tasks . . .

Hence, the market can be addressed in terms of groupings such as: large business, small business, state government, federal government, education, home, and personal services. These functional areas can be grouped separately or together, as their marketable solutions become better understood and implementable. Each market sector can be further differentiated by identifying the general industries within the grouping, and then again by denoting the specific products within each industry.

In actual implementation, products will span across functional areas. For example, a major market sector is large business. Within large business is the manufacturing industry, where the automobile is a key product. In looking at its more global implementation, the automobile industry will not only address the large business market sector, but will also use small business dealerships, parts stores, and repair shops. Together they all require extensive retail/wholesale distribution channels in order to achieve the automobile industry's unique position in the marketplace. (See FIG. 3-5.)

Types of users

As we consider the information users of each industry, they will be presented in terms of the tasks they perform, the type of network used in performing the task, and the specific type of information-handling performed. In this analysis, the types of information handling are defined as: inquiry/response (I/R), data collection (DC) (including polling and sensing), data distribution (DD), interactive remote access (RA), time sharing (TS), remote display (RD)/ documentation (D), interactive graphics (G), transactions (T), videophone/video conference (V/VC), and enhanced voice (EV)—such as, voice messages. (See FIG. 3-6.)

Therefore, types of information users are identified in terms of *industry's tasks—network types—type of information handling*; for example, in accessing patients' recorded status in the hospital from the doctor's office workstation, this would be described as a:

[narrowband network inquiry/response type of data user]

Type of Industry Task	Type of Network Service		
	Narrowband	Wideband	Broadband
————	Type of Information Need		
————	— — —	Type of Information Need	
————	— — —	— — —	Type of Information Need

Fig. 3-6. User-type model.

While the need to consult with other specialists on a series of X-rays requiring more bandwidth would make this communication a:

[wideband network remote display type of image user]

Information handling

So what are the various types of information? For the moment, let's consider the following:

- Inquiry/response (I/R): This inquiry/response data user makes a single or sequence of inquiries to determine the status of a remote file. The single inquiry will normally require fast connections and only hold for several seconds. The I/R user will, in general, operate on the publicly switched or privately leased narrowband network. The I/R call volume will be very large, as will be seen by the large list of I/R users for each industry and group.
- Data collection (DC): The data collection user collects data from remote sources by either initiating calls to poll or sense these sources or by receiving random data calls from them.
- Data distribution (DD): The data distribution user transmits in a broadcast mode to many remote terminals or distributes data to specific remote control facilities. This user may transmit large blocks of continuous bit rate (CBR) information or numerous bursty small messages of variable bit rate (VBR). The network operation has been over either leased lines of various grades of transport quality, or via the analog public switched network using numerous retransmissions to ensure transmission quality.
- Interactive remote access/time-sharing (TS): The remote access/time-sharing user requires a two-way transmission channel for data transfer. In remote access operation, large blocks of data are sent to and from centrally located computers with long idle times between transmissions. For time-sharing operation, short blocks of data are sent in interactive transmissions. Total holding time for time-sharing users will be in hours. Remote access users will usually require greater than POTS' (plain old telephone service) three minutes and less than an hour holding times per job.
- Remote documentation and display (RD): The remote documentation and display user will transmit and receive large data blocks of information, which usually require medium to high transmission speeds for a reasonable system input/output processing time. The complexity of documents or degree of resolution required dictates the volume of data to be sent in a given time frame, which in turn determines the transmission data rate. This type of user usually sends data blocks consisting of 16-bit to 24-bit pixel line scan information that's not necessarily specifically coded characters.

- Interactive graphics (G): This user accepts large blocks of data from a central computer to a remote computer, which refreshes a CRT display. The graphic user requires a large bandwidth to move these large data blocks to ensure a short "real time" system response.
- Transactions (T): This user requires a bidirectional data communication channel either for short data messages, such as teller-to-computer transactions, or for transferring large data blocks between computer systems. The user is characterized by machine-to-machine type conversion requirements, such as short system response and low error tolerance.
- Videophone/video conferences (V/VC): This video user will require an interactive facility to enable individual or group visual exchange of information. Here, dynamically available multiple transport channels will be required for more and more resolution, eye-to-eye contact, graphical/visual imagery to achieve "visual reality," and simulation of touch and move operations to enable remote physical contact, to obtain "virtual reality."
- Enhanced voice (EV): Numerous new features and capabilities are being added to increase the traditional black phone voice network offerings. As custom calling enabled call waiting, abbreviated dialing, so advanced custom calling, via the CLASS family features, will enable selective call waiting, selective call transfers, priority overrides, and selective blockage to take place as the telephone becomes an even more integral part of our daily lives. Here, message waiting, phones that announce the calling party by name, language translators, and phones that respond to verbal messages, will become commonplace.

Market analyses

It's time to further use the two models presented in FIGS. 3-5 and 3-6. As noted earlier, the information marketplace can be divided into major market sectors such as: business, education, government, residence, etc. These can be further subdivided (state government separated from federal government). We can then consider specific industries for each of these sectors or subsectors. In fact, some industries could be applicable to several sectors. In this manner, eighteen or so selected representative industries will be presented here, and each will then be reviewed in terms of its unique community of interest, its specific tasks, the type of information handling (I/R, etc.,) that's performed in the task, possible information services that could assist the application, and the types of users that perform the task in terms of their communication transport attributes. After analyzing these industries, we'll then, as FIG. 3-5 notes, consider general observations that reflect across all the industries—noting cross-industry information users, cross-industry information services, etc. This will lead to a conclusion of the layers of services that will be needed in the information marketplace, which will subsequently require the appropriate supporting network infrastructure that enables their delivery. Now, let's take a look at the industry applications.

Health-care industry

In exploring the health-care industry, it's interesting to consider the wide range of associated parties involved in health care. The community of interest configuration (FIG. 3-7) shows a highly integrated internal and external network of general practitioners, specialists, hospital administrators, technicians,

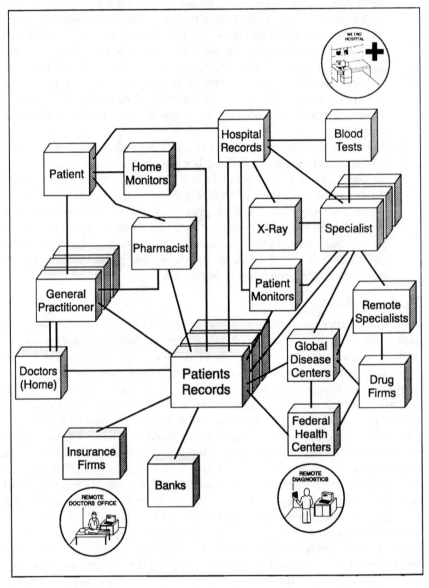

Fig. 3-7. Health care.

doctors, nurses, orderlies, clerks, researchers, testers, and equipment suppliers that are supported by federal and state statistical data centers, inventory control mechanisms (equipment, drugs, bandages), and record archives.

These participants are not only internal to one particular hospital or within one HMO, but are also interconnected throughout the area to multiple HMOs, doctors' offices, hospitals, research centers, specialists, pharmacies, poison control centers, state and federal (national) agencies, as well as insurance firms, banks, and funeral homes.

As we consider both an internal view from a hospital or medical center perspective as well as an external view (which extends to small businesses within the local community and to large state or federal databases), we need to next analyze the specific tasks that have become an integral part of the medical community's mode of operation. As better communications are established, these needs will change. We'll consider these activities in terms of an easy-to-understand model.

Tasks

The tasks in TABLE 3-1 show an interesting, comprehensive view.

Table 3-1. Health-care Tasks

	Narrowband	Wideband	Broadband
Hospital patient files status/medicine	I/R		
Insurance-hospital claims	RD		
Remote documented orders	RD		
Hospital med. records	T		
Hospital eqp. records	T		
Dr.-Dr. communications	EV-RD	RD-G	V/VC
Remote Dr.-patient communication	EV-DC		V
Dr.-education (remote) files	RD-DD	RD-DD	V/VC
Dr.-consultation files	EV-RD	RD-DD	V/VC
Testing/lab reports	DC		
Remote test reports	DC		
Remote test monitors	DC		
Prescription drugs	RD-T		
Blood tests	DC		
X-rays	V-RD	V/VC	V/VC
AIDS tests	DC-DD		
Blood banks orders	I/R-T		
Poison centers inquiries	IR		
Accident reports	RD		
Artificial limbs	I/R-T		
Kidney banks	I/R-T		
Eye banks	I/R-T		
Autopsy reports	RD		
Contagious disease center alerts	DD		

Information services

As we look at the amounts and types of inquiries, responses, data sensing, data collection, data distribution, remote documentation and display, and visual communications, we see the need for the services listed in TABLE 3-2.

Table 3-2. Health-care Information Services

• Transport computer interfaces	• Global access
• Message switching	• Record archiving
• Video education	• Patient sensors
a. On demand	• Home monitors
b. Broadcast	• E-mail
• Video conference centers	• Voice mail
• High-speed transport	• Video mail
−Point to point	• Personal computer terminals
−Point to multipoint	• Wideband X-ray network (home, office)
• Video/imaging	• Information switching
• Graphic/X-ray-archives	• Multiple database access

Information users

By overlaying narrowband, wideband, and broadband forms of communication on these tasks, we may better understand and appreciate the advantages (and disadvantages) of the various communication (and computer) alternatives. Here we see that a key attraction to the health-care community would be fast-connect narrowband (64K/128K) data-handling network that enables test results, sensors, monitors, patient records, insurance forms, disease alerts, and equipment orders to flow here and there—accurately—securely, and without delay.

As information becomes more graphic and data intensive, there's also a need for the wideband networks. Once the resolution requirements become acute and the need for visual conferences and discussions occur, we've moved to the broadband arena. Here, connect time is not as important as having access to long-holding-time, high-capacity channels. Error correction mechanisms can help offset noise as compression techniques are less sensitive with higher bandwidth. Educational "on demand" video or "broadcast" video facilities will also be needed.

Therefore, the overall needs of these users may be viewed as shown in TABLE 3-3.

Table 3-3. Health-care Narrowband-Wideband-Broadband Users

Narrowband	(70%)*	IR-DC-DD	Public switched data users
Wideband	(25%)	TS-RD	Internal LAN/Public switched data/image users
Broadband	(5%)	VC	Public switched/internal LAN/external MAN users

*% usage of overall attempts.

General observations

As could be expected, the medical industry's users cover the full range of narrowband, wideband, and broadband services. However, on closer look, it's interesting to note that a highly desirable and used facility would be a fast-connect, narrowband, data-switched public network with appropriate mechanics for enabling dissimilar computers to interconnect to obtain interprocessing via private-to-public internetworking interfaces. This could be deployed via existing facilities, but it must be ubiquitous. This is key and essential to promoting extensive usage.

There's also the need for graphics, X-rays, and other imaging to be transported. Some CPE systems can generate as many as 100 or so images in still frame or even in full color video. Thus, there's an expanding need for wideband and broadband transport. Some dynamic capacity mechanisms can provide relief by using multiple narrowband facilities in parallel. As additional lines are used to carry information, as needed, the information can be reassembled sequentially, at the end, before delivery. This then leads to frame-relay-type wideband capabilities for a variable number of transport channels. Finally, most educational needs will require full video capabilities, as doctors review new operating techniques or monitor online remote operations.

As an aside, we cannot underestimate the importance of success. As more and more services go "online," as more and more doctors and nurses, administrators, pharmacists and test centers communicate in "real time," the network cannot be designed on a contention basis in which users share a limited availability of capacity. (It must be available 100 percent of the time. It can't go down or be under limitations.) Each user will need their access unlimited as much as possible. A successful network cannot introduce delay. Therefore, narrowband transport mechanisms cannot cause undue input/output delays to processing mechanisms. As these develop, the shift must be feasible to high-speed transport systems that can more successfully and quite economically handle the greater demands of higher volume and higher resolution.

Large business, manufacturing industry, small business, wholesale/retail trade industry, automobile industry

If one industry were selected to represent the complete flow of raw material in/product out, product to the showroom, product in the customer's home, it would be the automobile industry. It's the only industry in America that clearly spans many major market sectors and industry groups—large and small business, manufacturing, wholesale and retail. Each set of unique tasks for each industry has to be integrated in the overall flow from customer order to product delivery, to product operation, to product repair and support, to product termination. This delivery process is truly global in scope and universally similar in execution. It may be academically more complete to consider each industry separately, but this particular application is so keenly understood that it will be fully analyzed here to show not only each industry's major functions, but the also interdependencies of cross-industry tasks that require cross-industry network and service-integrated solutions. So let's assume for the moment that we'll have the customer needs and requirements input from the small-business car dealerships and take a brief look at the overall manufacturing process.

Manufacturing

The manufacturing plants for today's automobile are one part manufacturing, one part assembly, with an administrative part coordinating the order, manufacturing, and arrival of the "other parts" that make up the automobile. (In reviewing how foreign cars are "assembled" in America, some might say that the coordination and assembly of foreign parts is now the new definition of "made in America.")

We no longer need yesterday's plants. It's a new world with new needs and requirements. As in the manufacturing of communication switching systems, no longer are we able to have raw material enter one door and have switching parts go out the other side, as did the GTE Automatic Electric one-mile-square "step switching" plant of Northlake, IL, or the Western Electric Hawthorne "#5 Crossbar" plant in Chicago, where angle iron was bent to make equipment frames and various spools of copper, gold, silver, aluminum, and plastics were melted down and stamped out to form reeds, relays, printed circuit boards, and wire harnesses. No longer are telcos told to buy this or that product from their own suppliers. Similarly, in the automobile industry, no longer are prices set at arbitrary levels with little competition. No longer are products designed with limited concern for the customer's usage convenience. Today's and tomorrow's manufacturing plants are becoming increasingly flexible to consumers' customized needs.

There are now sophisticated marketing, purchasing, design, and testing departments to support manufacturing. Every operation today has computers "online." There was a time in the early '60s that computers were only there to support payroll and perform a limited set of work for the cost-accounting department. Now, computers control inventory flow, tool-and-die setups, robotic welders, simulated and actual parts testing and "burn in," outside job shop orders and deliveries, product design, experimental design, research of different materials, and the prototyping of new models.

Distributed arrays of computers now share job order information and work flow, to ensure "just in time" deliveries occur on the line "in time," having the right part, the right color, the right car for the right customer. This requires an extensive internetworking of order information, material design plans, construction setups, delivery scheduling, testing, and assembly coordination. From this world came the LANs using GM's internetworking manufacturing protocols, MAPS, and Boeing's internetworking design protocols, TOPS, as streamlining the production line became the prime direction in saving time and money. (See FIG. 3-8.)

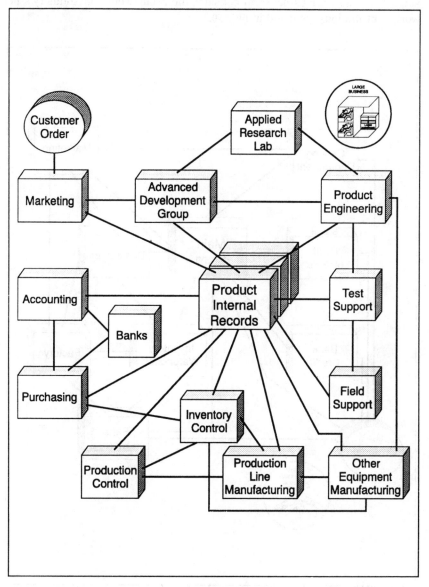

Fig. 3-8. Manufacturing plant.

Retail/wholesale

Driving these manufacturing plants is linkage to their marketing groups. This is the complete network of dealerships, privately owned by small businesses, who rely on the extensive distribution networks that are so key to the dealerships' success. So, let's pause to consider the retail/wholesale industry's network contributions, as noted in FIG. 3-9.

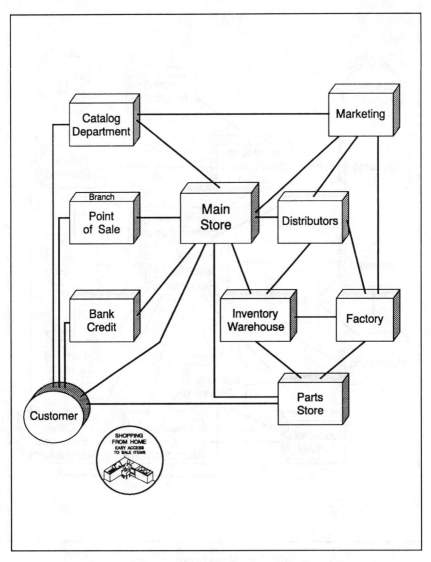

Fig. 3-9 Retail trade.

The backbone of these industries are the computer control systems that facilitate the movement of products from manufacturing plants to regional or state distribution warehouses, from which they're distributed to local distribution centers or directly to the main store or branch stores. One challenge is to bypass as many holding locations as possible and ship the product directly from the factory to the point of sale, or at least to a local-serving-area supply center. *Shelf life* or *lot life*, the period that the product remains unsold in the possession of middlemen or store owners, is a measure of lost revenue. This is the period the banks make money at 8 to 16 percent for borrowed financing money to build the product or "stock the store." Similarly, the carrier costs for shipping products between stores must be considered to achieve the right balance in the store inventory-warehouse trade-offs.

Small-medium business

In todays' competitive arena, distribution networks are essential to small business in order for their weekly orders to be properly and quickly filled to re-stock depleting inventories. Also, in the same sense that communication is essential for store managers to internetwork with their suppliers, it has also become a key factor in enabling customers to select their goods from remote home, office, or construction-site locations. Here, inquiry mechanisms are being developed to enable remote browsing of store items, reviewing store sales, ordering lists of items, monitoring status of unfilled past orders, etc.

Automotive industry

With these considerations in mind, FIG. 3-10 denotes the far reaches and cross avenues of the automotive industry. The industry path proceeds from the small business sales order to the manufacturer, through the retail/wholesale distribution network, to delivery of the finished product to the customer. Then the industry is supported by repair shops and maintenance shops for replacement parts and services.

As banks and insurance firms are integrated into the complete picture, together with gasoline stations, car radio shops, tire dealers, state testing, accident and auto theft departments, and the department of roads and highways, we begin to see the scope and magnitude of the need for ubiquitous data networks to interconnect the diverse and remote operations of the total automobile industry complex. (See FIG. 3-11.)

Tasks

A brief look at the functions performed during daily operations easily can show the need for diverse networks to enable diverse tasks to be more effectively accomplished. (See TABLE 3-4.)

Table 3-4. User Tasks and Transports

Manufacturing	Narrowband	Wideband	Broadband
Dealer orders	T		
Internal corporate messages	DD		
Inventory control	I/R-DC-DD		
Planning	I/R-DC		
Marketing	I/R-DC		V/VC
Purchasing	I/R-T		
Personnel	I/R-DC-DD		V/VC
Legal	I/R-DC		
Logistics	DC-DD		
Field maintenance	DD-RD		
Design-hardware	TS-RD	TS	V/VC
Design-software	TS-RD	TS	V/VC
Design-support-facilities	DC-RD	TS	V/VC
Accounting	I/R-DC		
Quality tests	DC-DD		
Parts tests	DC-DD		
Performance tests	DC-DD		

Retail/wholesale trade	Narrowband	Wideband	Broadband
Product orders	T		
Credit ref. checks	I/R		
Central distribution	DD		
Store-to-store inventory control	DC-DD		
Central billing	T		
Management info. exchange	I/R-DD	RD	
Bank-retail transactions	T		
Mail-order business	I/R-T		
Freight handling	T		
Returns control	T		

Small/medium business	Narrowband	Wideband	Broadband
Customer-orders	T		
Point-of-sale transactions	T		
Credit checks	I/R		
Inventory control	T		
Auto rental chains	I/R-T		
Education/sales information	RD	RD	V
Food industry distribution	I/R-DC-DD		
Food distribution for restaurant chains	I/R-DC-DD		
Drug business order/customer bill	I/R-T		
Inventory control for chain stores	DC-DD		
Hotel/motel reservations	I/R-T		V

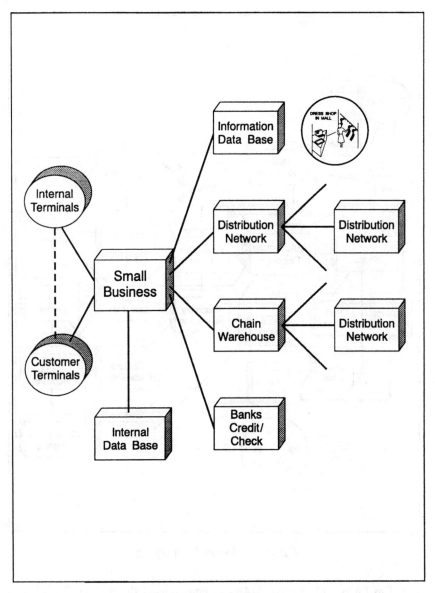

Fig. 3-10. Small business.

Information users

In reviewing the types of tasks and the kinds of narrowband/wideband data-handling networks available today, it's quite apparent that due to the absence of narrowband data-switched facilities, internal information networks have

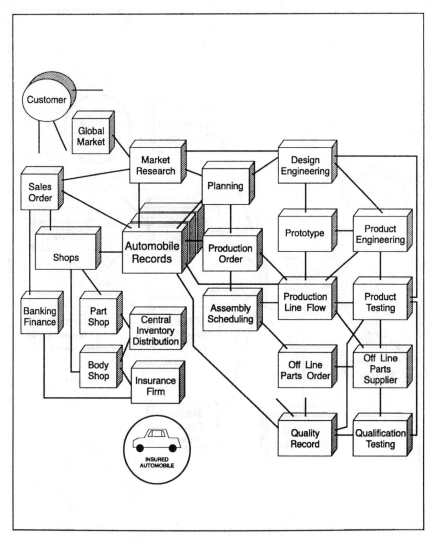

Fig. 3-11. Automobile industry.

turned to LANs and external MANs to interconnect the proliferation of small LANs, as well as to external voice grade modems for dial-up access and transport. However, as information handling becomes more ubiquitous and increases in terms of visual and graphic information, a switched fiber network will become more and more desirable. Externally, it remains a world of small transactions, inquiry responses, customer requests, automobile inventory control, multiple manufacturers parts control and delivery, and change orders that fit nicely into the narrowband offerings handled by a public data switched network.

In time, as more visual information is actually updated in local dealerships, manufacturing communications needs could increase from the narrowband rate to wideband. Alternatively, longer duration broadcasts at narrowband rates could be sent in nonbusiness hours overnight and stored in local dealer computers. Here, users appear to fall into two groups. Externally, narrowband will be needed for store owners, with additional wideband for some wholesalers and dealers. Internally, narrowband, wideband, and broadband will be required for manufacturing. In time, full broadband fiber will merge and integrate these internal operations. (See TABLE 3-5.)

Table 3-5. Manufacturing Narrowband-Wideband-Broadband Users

Internal	Narrowband 30%	Remote display	Data/imageusers
	Wideband 40%	Data transfer	Data/image users
	Broadband 30%	Video display	Image/graphic/video users
External	Narrowband 90%	I/R-DC-DD	Data/image users
	Wideband 10%	RD-video display	Image/video users

Some of the information services include:

- Customer order.
- Point of sale.
- Inventory control.
- Customer requirements-assembly orders progress.
- Credit check access programs.
- Bank-loan access programs.
- Billing programs.
- E-mail.
- Video-sales brochures.
- Graphic design tools.
- Inventory/assembly control tools.
- CAD/CAM.
- Service monitoring.
- Order flow transaction network.
- Product locator programs.
- Modification-retrofit change programs.
- Modeling and simulation.
- Test vehicles.

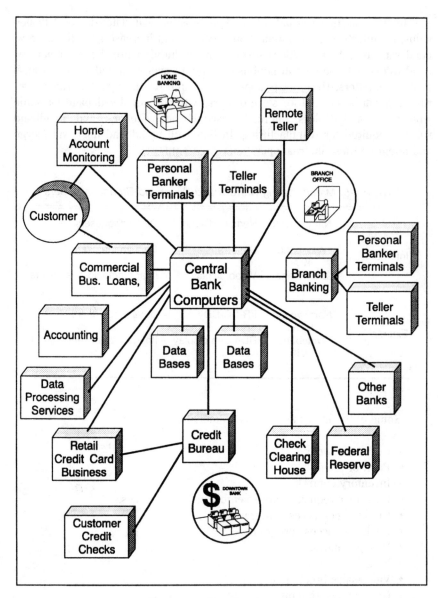

Fig. 3-12. Banking industry.

Banking/financial industry

Today's banks are becoming bigger and bigger, as mergers and acquisitions are expanding base operations, and as states are reassessing branch banking rules and multiple services regulations. The financial problems of the early '90s have hastened internal automation of operations, as many of the industry's tasks are being centralized. This is even occurring in group and family banks that previously had quite separate operations.

The savings-and-loan and banking crises have caused a trail of bank-ruptcies and failures that have shown the negligence and ineptness of the federal bank inspectors. It has caused a considerable scare in many who once believed the failures of the Great Depression could not be repeated. It has caused a considerable increase (100 + billion?) in our national debt. No longer will the industry be the same. It will be watched and monitored. It will be merged. It will change and change again in order to be more efficient and more effective. (See FIG. 3-12.) All in all, this provides further pressure on banks to offer full money market services, as they automate every aspect of their internal operations, using computers and communications.

Tasks

Every task within the banking community is being changed to speed up money handling, credit checks, and new services that enable banks to be more competitive in the marketplace. (See TABLE 3- 6.)

Table 3-6. Banking Tasks

	Narrowband	Wideband	Broadband
Teller trans.	T		
Remote teller trans.	T		
Credit verification	I/R		
Check clearing	T	T	T
Remote branch bank trans.	I/R-T		
Remote account monitoring—bus	I/R-T-EV		
Remote account monitoring—home	I/R-T-EV		
Federal reserve trans.	I/R-DC	T-DD	T-RD/D
Bank-to-bank trans.		T	T
Bank DP services	TS	TS-RD	TS
Retail-Bank automatic trans.	TS-T	T	T
Checkless Trans. (credit cards)	TS-T	T	T
Home banking	I/R-T		V
Education	EV		V/VC
Internal conferences	EV		V/VC

Information users

In reviewing the network solutions to meeting the internal and external needs of the various banking tasks, it's quite evident that an internal network that doesn't inhibit or delay the numerous inquiries and transactions to personal accounts is essential. This is true whether these inquiries are from the main tellers, private tellers, or one's home. Secondly, there's a need for higher speed transports to handle the higher volume in transactions between banks and the Federal Reserve and check clearinghouses. Finally, the world of credit card transactions has blossomed into a major retail-bank parallel network.

Later, as banks continue to locate closer and closer to the customers, there are visions of no central banks. There would simply be small offices here, there, and everywhere adjacent to shops, restaurants, and airline terminals. The key to banking will be access at any time, from any place, as new hours include evening videophone conversations with personal bankers as well as full service dial-up dataphone services from the home. The interim information users are listed in TABLE 3-7.

Information services

- Home banking.
  a. Account inquiries and changes.
  b. Automatic deposits.
  c. Automatic bill payment.
  d. Check classification grouping—year end.
- Data network—transactions.
- Money market investments.
- Federal Reserve transactions.
- Account assessments—graphical displays.
- Home mortgage programs.
- Commercial loan programs

Table 3-7. Banking Narrowband-
Wideband-Broadband Users

Narrowband 30%	Inquiry/response	Voice/data/image users
Narrowband 50%	Transaction	Data users
Wideband 15%	Transaction	Data users
Broadband 5%	Videophone	Video users

Securities and investments industry

Stockbrokers, financial advisors, commodity traders, bond sellers, arbitrators, and real estate sales workers, among others, are becoming networked together as New York's big board is tied to the futures market and as traders play both the shorts and longs to hedge their "bets." The market is becoming global as off-hour transactions are allowed and as foreign markets are accessed.

A major reason for the "big scare" 500-point drop in '87 was the overload caused by programmed trading. The private exchange networks could not handle the volume, as markets became more and more delayed and out of sync. Postmortem analyses have clearly showed how dependent the entire securities and exchange markets have become on "real time" response. This is especially true as market volume increases.

Figure 3-13 shows the complexity of networking all the entities. Business will change to better accommodate the customer and enable direct access for home or business transactions under customer control and provide more and more information to the customer anywhere, anytime. Hence, there will be a

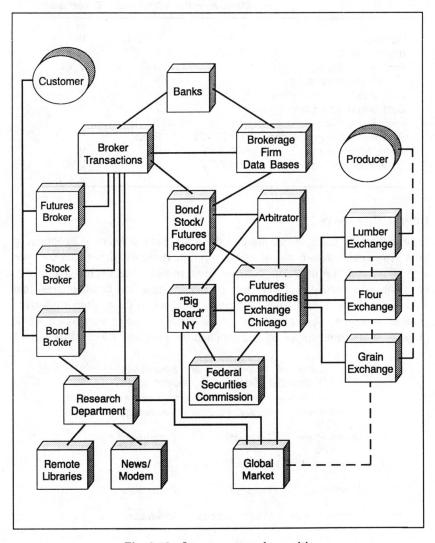

Fig. 3-13. Investments and securities.

need to reach out further and further with more and more information as visual communications, eye-to-eye, will become a key sales tool that is worthy of investment.

Tasks

TABLE 3-8 shows the heavy emphasis on inquiry/response and transaction uses.

Table 3-8. Investments and Securities Tasks

	Narrowband	Wideband	Broadband
Inquiries	I/R-EV	RD/D-G	V
Buy/sell trans.	T-EV	T	
Portfolio monitoring	I/R	G	V
Remote-to-central trans.		T	T
Central-to-stock exch. trans.		T	T
Central-to-remote billing		T	T
Internal paperwork trans.		T	T
Research survey/monitoring	I/R-DC-DD	RD/D	
Home Access/trans.	I/R-T-EV	G	V
Office Access/trans.	I/R-T-EV	G	V/VC

Information users

In reviewing the tasks, it appears that the main types of users will continue to be narrowband between customer and broker. However, depending on overload conditions, the networks between brokers and stock exchanges need to be wideband or broadband to ensure that the volume can be adequately handled. Videophone remains the least-understood tool for the customer interface. How often would eye-to-eye contact have made the sale? How often would graphic charts have supported and clinched the decision? (See TABLE 3-9.)

Table 3-9. Investments and Securities
Narrowband-Wideband-Broadband Users

Narrowband 40%	Inquiry/response	Data/image users
Narrowband 30%	Transaction	Data users
Wideband 25%	Transaction	Data users
Broadband (alt) 25%	Transaction	Data users
Broadband 5%	Inquiry/response	Video users

Information services

- E-mail.
- V-messaging.
- Videophone.
- Security.
- Reliability.
- Integrity.
- Graphic packages.
- Future projections.
- Financial analysis.
- Remote documentation.

Insurance industry

Seven out of ten insurance agents work out of their home, with a personal computer that can access the firm's "canned programs" for various policies. They simply feed the program the particulars, it executes, and then remotely prints the information at the PC location, or they pick up their computer "runs" at their branch office. Each major city has a main office. There a couple of floors of bull-pen offices, with terminals at each desk and a phone, enable workers to carry on a conversation with a potential applicant. At the same time, the employee can obtain intermittent/interactive access to the mainframe for policy information. Hence, the mode of operation is either among office mainframe computers or between terminals and computers in the office, or to and from PCs at home and their appropriate office mainframes. (See FIG. 3-14.)

Tasks

A list of tasks in the insurance industry is shown in TABLE 3-10.

Table 3-10. Insurance Tasks

Insurance (car, house, life, health)	Narrowband	Wideband	Broadband
Credit ref.	I/R		
Policy transactions	T		
Remote inquiries—customer	I/R		
Time-sharing business for DP	TS		
Claims	T		
Policy payments and billing	T		
Home mortgage	TS		
Home sales	I/R-T	RD/D	
CPU-CPU transactions		DC-DD	

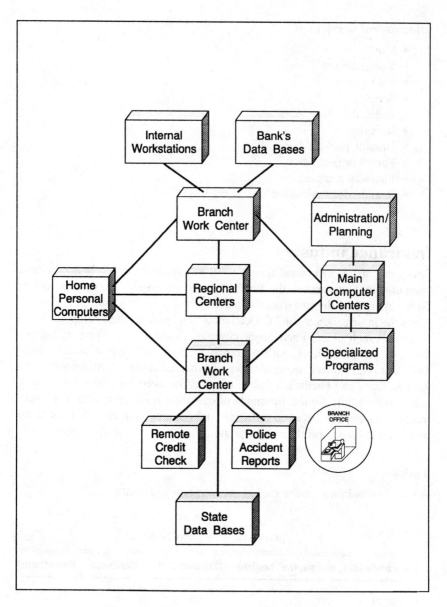

Fig. 3-14. Insurance (car, house, life, health).

Information users

The insurance industry is specifically tied to interactive computing over narrowband facilities. Mainframes are provided "specifics" for their various "runs," for which they deliver alternative inputs. Therefore, there are three typical users, as shown in TABLE 3-11.

Table 3-11. Insurance Narrowband-Wideband-Broadband Users

Narrowband	40°	Inquiry/response	Data users
Narrowband	40%	Transaction	Data users
Wideband	20%	Data distribution—data collection	Data users

Information services

- Insurance programs.
- Home mortgage programs.
- Home-listing real estate.
- Remote documentation.

Police, protection, law enforcement industry

This industry is heavily inquiry/response oriented. Access can be made via mobile link to the local area station, which is connected via wideband links to "central." At "central," there exists not only a local mainframe, but also a gateway system capable of accessing numerous additional databases listing various types of information that, for example, aid law enforcement and protection investigations. Their local complex system is one in which front-end processing is performed at gateways. Calls are then routed to the appropriate alternative internal computer systems or returned to the network to be routed to remote destinations. Data collection takes place by launching a series of calls to distant databases to inquire for more information. (See FIG. 3-15.)

Tasks

Police, FBI, state, and federal law enforcement agencies have numerous files that are constantly prepared, updated, and distributed. (See TABLE 3-12.)

Table 3-12. Law-Enforcement Tasks

	Narrowband	Wideband	Broadband
Crime monitoring	I/R-DC	DD	V
Criminal files & ID	I/R-DC	DD-G	V
Auto thefts/files/accidents	I/R-DC	DD	
Drug files	I/R-DC	DD	
Agency communication	I/R-DC	DD	V
Licenses files	I/R-DC	DD	
Federal files	I/R-DC	DD	V
Missing persons, theft, etc.	I/R-DC	DD-G	V
Alarm monitoring	DC-DD		V

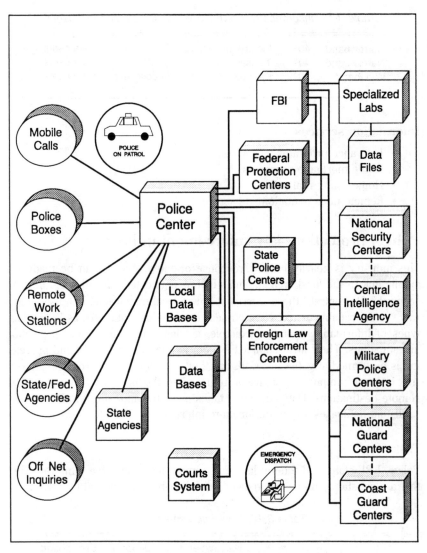

Fig. 3-15. Law enforcement protection agencies.

Information users

Users are putting information into their particular database, searching databases for additional information, or scheduling operational activities for personnel and equipment. High-speed links between larger offices and state and federal databases help speed up the overall operations. Video communications would assist in many areas. (See TABLE 3-13.)

Table 3-13. Law-Enforcement
Narrowband-Wideband-Broadband Users

Narrowband	25%	Inquiry/response	Data users
Narrowband	25%	Data collection	Data users
Wideband	25%	Data distribution	Data users
Broadband	25%	Video display	Video users

Information services

- Browse mechanisms.
- Search mechanisms.
- List-processing programs.
- Database management programs.
- Graphical imaging programs.
- Video filling/archiving/display.

Government—state/city industry

State government and their city municipals must deal with many social problems resulting from years of inefficiently funded and managed programs on both the state and federal level. As drugs invaded the neighborhoods, as jobs declined and population increased, as children were left to street survival, as the gangs of Los Angeles and Chicago migrated to Denver, Seattle, Minneapolis, etc., as people emulated television's violence and sex, as schools provided inadequate basic education, etc., etc., the result in the '90s is tremendous pressure on the state and city governments to deal with highly complex social issues in order to reestablish a functioning society that provides opportunity for a better quality of life. Population growth and moral decline further complicate the problems.

To meet such challenges, every facet of the governmental structure is becoming "databased," as every department goes "online," requesting the ability to access every other department's databases. States are turning to private overlay networks to interconnect their communication infrastructure to handle their expanding information exchange. Many of their more remote or autonomous operations request access to centralized databases via public networks. Hence, the '90s is seeing a mix and match of private and public networking and internetworking. (See FIG. 3-16.)

Tasks

Note the information listed in TABLE 3-14.

Table 3-14. City-State Government Tasks

	Narrowband	Wideband	Broadband
Operation communications	I/R-TS	T-RD	VC
Large city communications	I/R-TS	T-RD	VC
Taxes (wage, real estate, sales, etc.)	T-DC	T	
Health and welfare trans.	I/R-DC	T-RD	
City and county (birth, death, disease)	I/R-DC	T-RD	
Legislation operation	I/R-TS	T-RD	VC
Judicial operations	I/R-TS	T-RD	VC
Road construction/maintenance	I/R-DC	T-RD-G	
Licenses	I/R-DC	T	

Information users

Inquiry/response, database collections, transactions, and remote documentation are key elements of operation as users require time-sharing, database management, inquiry, browse, and search mechanisms. A government is run on statistics; at the state and city level, governments are beginning to look and act like their federal counterparts. There are several types of users listed in TABLE 3-15.)

Table 3-15. City-State Government Narrowband-Wideband-Broadband Users

Narrowband	30%	Inquiry/response	Data users
Narrowband	30%	Data collection	Data/image users
Wideband	30%	Transaction-documentation	Data/image users
Broadband	10%	Video conferencing	Video users

Information services

- Scheduling programs.
- Inquiry/response programs.
- Video conferences.
- Data-handling network services.
- Browse/search programs.
- Database management.
- Protocol conversions.
- P&P internetworking.

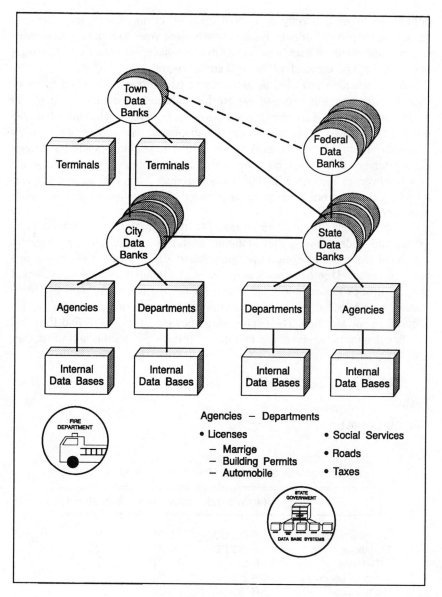

Fig. 3-16. City/state government agencies.

Federal government

Federal government grew and grew until gruesome, as one in every three citizens is now directly employed by the government and one in two citizens is either directly or indirectly dependent upon government, city, state, federal, military, or research budgets. Each new president vows to reduce government size and spending, as each new administrator is faced with an ever-increasing debt—at the 4.05 trillion level in 1993 and growing.

As information provided in government publications, furnished to every citizen at government expense, so quickly indicates, one major role of the federal bureaucracy is to provide statistics—not necessary solutions—but documents on what to do, how to do, and statistics, statistics, statistics on what was done. Managers of the government's internetworking needs have been hoping for relief. The federal government is attempting to develop new national networks, such as FTS-2000, to help reduce communication costs and speed up access from their growing number of computers to their growing number of databases.

Unfortunately, protocol conversion to their numerous dissimilar computers and addressing to their voluminous LANs, as well as new extensive data-handling requirements, have complicated their private voice-based network offerings. This has encouraged many of the agencies and the different military groups to continue their search for more efficient and effective information networks, while enabling access to their internal networks via unsecured public facilities. Their quest continues over the '90s to find the most functional information-handling mixture of private and public networks. (See FIG. 3-17.)

Tasks

(See TABLE 3-16.)

Table 3-16. Federal Government Tasks

	Narrowband	Wideband	Broadband
Legislation	I/R-DC-DD	TS-RD	VC
Judicial	I/R-DC-DD	TS-RD	VC
Defense	I/R-DC-DD	TS-RD	VC
Social security	I/R-DC-DD	TS-RD	
Statistics	I/R-DC-DD	TS-RD	
Internal revenue	I/R-DC	T	
Health and welfare	I/R-DC-TS	RD	
Patent	I/R-DC	RD	
Library	I/R-DC	RD	
Federal-state info.	I/R-DC	RD-G	

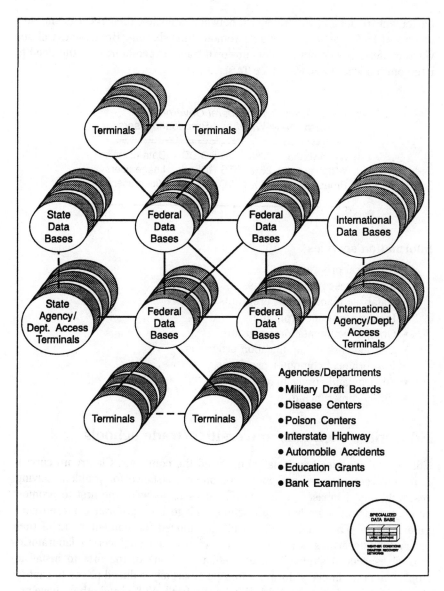

Fig. 3-17. Federal government—typical data collection/distribution network.

Information users

The agencies are just beginning to internet data; they're naturally more voice oriented. Hence, narrowband will see extensive I/R, DC, and DD information movement. Wideband will transport, between agencies, their records and documents (RD/D), as well as enable remote time-sharing. Broadband will see video conference centers to help cut down travel expenditures, as the need to intercommunicate increases. (See TABLE 3-17.)

Table 3-17. Federal Government
Narrowband-Wideband-Broadband Users

Narrowband	70%	I/R-DC-DD	Data users
Wideband	25%	TS-RD	Image users
Broadband	5%	VC	Video users

Information services

- Statistical programs.
- Inquiry/response programs.
- Database management programs.
- Imaging/graphic programs.
- Online documentation programs.
- Internetwork communication programs.
- Interprotocol computer programs.
- Security and survivability.
- Video conferencing.

Education—schools/universities/trade schools

Educational facilities have finally embraced the computer. Classroom curriculums will require more and more computer assistance for problem solving, research, and databased references. Trade schools were the first to promote computer instruction; soon high school college prep programs offered computer classes. Then, universities gradually moved the computer out of their accounting departments and computer science and engineering laboratories into all aspects of study. Carnegie-Mellon was one of the first to install an Apple computer in every dorm room. Apple partially reduced prices to capture the students' pocketbooks. IBM, Hewlett Packard, AT&T, and others have provided discounted clusters of terminals for student workshops; while DEC and Wang concentrated on group offerings to the business firms' training departments and word-processing organizations. Indeed, the computer has penetrated every aspect of the Engineering and Science departments, as well as the economics, financial, and business schools.

Computers now assist in the daily operations of the colleges and universities, although many in the liberal-arts colleges haven't yet found much use

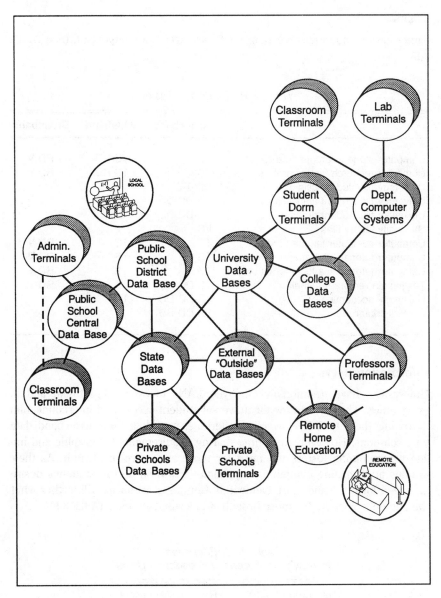

Fig. 3-18. Education.

for the computer other than word processing and browsing through book reviews or subject searches. Remote visual training provides both "online" and
"offline" access to classroom presentations. The university environment is especially good for innovative software programs for every aspect of learning,
from financial analysis to games. (See FIG. 3-18.)

Tasks

Time-sharing and inquiry/response database access are important. (See TABLE 3-18.)

Table 3-18. Education Tasks

	Narrowband	Wideband	Broadband
Computer-student remote time share	TS	TS-RD	RD-V
Computer-research faculty prob. sol.	TS	TS-RD	RD-V
University file/student bills	DD-T		
University file/student statistics	I/R-DC-DD		
University file/student grading	I/R-DC-DD		
University library files	I/R		
Computer-computer language share	TS	TS	RD-V
Computer-tutorials via remote terminals	TS	TS	RD-V
Computer-tutorials via local terminals	TS	TS	RD-V
Testing via remote terminals	TS-DC	TS	RD-V
University accounting	I/R-DC-DD	T	
Book distributors	I/R-DD-RD		

Information users

Universities have traditionally deployed LANs in every department and in every office. Their need now for universal student access and interconnection is driving them to ISDN switched offerings, providing the narrowband data type solutions. However, their time-sharing requirements for graphic and image programming fosters the need for more and more bandwidth. As their graphic and imaging computing capabilities grow, so do the student's desire and need for more and more bandwidth. Computer art forms will further whet the students' appetite for more broadband capabilities. (See TABLE 3-19.)

Table 3-19. Education Narrowband-Wideband-Broadband Users

Narrowband	10%	I/R	Data users
Narrowband	10%	DC-DD	Data users
Narrowband	50%	TS	Data users
Wideband	20%	TS	Image users
Broadband	10%	V-RD	Video users

Information services

- Financial programs.
- Economic programs.
- Engineering design programs.
- Architectural/statistical programs.
- Graphics programs.
- Games.

Home communications

As the telephone encompasses the dataphone and changes to videophone, more and more information capabilities will be accessible from the home. By the early '90s, what was once a fad industry has become a major contributor to the gross national product—as those who work at home have become 22 percent of the labor force. Many disillusioned workers of the '80s have seen their families torn apart with child-care problems, drug problems, marriage problems, and personal exhaustion. Moms could not be successful supermoms while attempting to hold both family and job together. Many have turned to working out of their homes, using personal computers and computer communications, to achieve both personal satisfaction and family success.

These expanding communication networks will continue to provide exciting services to the home, as visual capabilities enable remote computing, remote shopping, remote education, home banking, and interactive videophone, as well as high-definition television news, sports, and entertainment programs. In addition, newspapers, letters, and magazines will be transmitted electronically. Monitoring and control of home alarms, energy management, and meter reading of home appliances can be obtained from the local utilities. All this and more can be achieved by providing access to and from the home to various central computer systems and distributed databases. (See FIG. 3-19.)

Tasks

Jumping from narrowband to broadband—an either/or situation. (See TABLE 3-20.)

Information users

In addition to high quality and enhanced voice services such as voice messaging and CLASS, the full range of narrowband services, from time-sharing to data collection, data distribution, inquiry/response, and basic transactions are accessible to the home. In the future, broadband interfaces will be needed for educational and entertainment distribution services, as well as videophone.

Table 3-20. Home Communications Tasks

	Narrowband	Wideband	Broadband
Voice communication	EV		
Time-sharing accounts management	TS		
Shopping selections, remote	I/R		V
Small student calculations	TS		
Newspaper bulletins	DD		
Mail electronics	DD		
Remote work	TS		V
Remote education	TS		V
Home incarceration	DC		
Alarm monitoring	DC		
Meter reading	DC		
Energy management	T		
Entertainment			V
Video communication			V

Information services

- Alarm monitoring.
- Energy management.
- Meter reading.
- Dataphone.
- Videophone.
- E-mail.
- V-mail.
- Remote education.
- Database access.
- Voice messages.
- 2nd line
- CLASS services.

Utilities

Operating companies, such as telephone, power, and gas, need extensive communications to aid their distributed operations. Parallel data message networks transport control and status information around their facilities to provide needed signalling to cause alternate route selection, load control, reconfiguration, shut down, and restart functions, as well as numerous throughput measurements and status reports. The work force also needs massive order con-

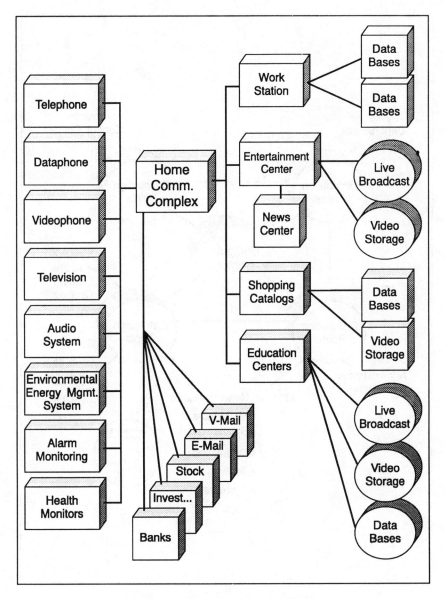

Fig. 3-19. Home communications center.

trol, vehicle control, equipment control, and remote test/support systems to enable central and remote distribution channels to function synchronously to satisfy customer needs and resolve system failures in a timely fashion. (See FIG. 3-20.)

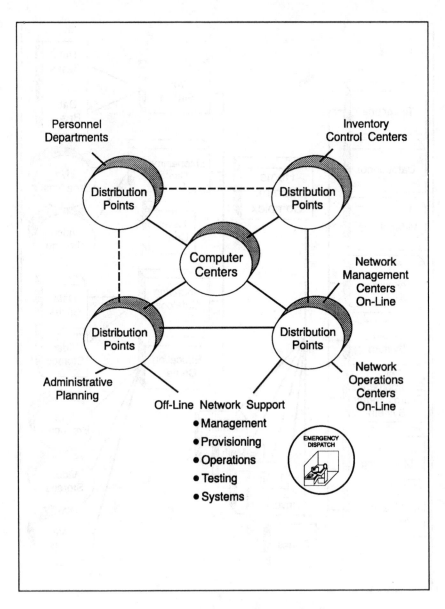

Fig. 3-20. Utilities.

Tasks

For utilities, there's considerable data movement and transfer. (See TABLE 3-21.)

Table 3-21. Utility tasks

Operations (utility, oil, tel, pwr, gs.)	Narrowband	Wideband	Broadband
Distribution control	I/R-DC-DD	T-TS	
Planning	I/R-DC-DD-EV	T-TS-G	V/VC
Records/statistics	I/R-DC-DD	T-TS	
Maintenance/operations/provisioning	I/R-DC-DD-EV	T-TS-G	V/VC
General management administration	I/R-DC-DD	T-TS	VC

Information users

Besides the narrowband voice services, there will be considerable data handling in the form of inquiry/response, data collection, data distribution, within each remote distribution complex, with wideband facilities and eventually broadband to enable better communications between distribution points, central control, and administration nodes. Extensive status and monitoring information is needed for future planning and expenditures. There's also the need for video conferencing around the network between work centers. (See TABLE 3-22.)

Table 3-22. Utility Narrowband-
Wideband-Broadband Users

Narrowband	80%	I/R-DC-DD	Data users
Wideband	15%	T-TS	Data users
Broadband	5%	Video conference	Video users

Information services

- Data-handling transport.
- Data access and storage programs.
- Order control programs.
- Message systems.
- Alarm monitoring systems.
- Video conferencing systems.
- Voice messaging systems.
- Closed user group services.

Transportation

Transportation firms, such as the airlines, are under intense financial pressure to streamline operations and appropriately communicate status and availability to their customers and their sales agents. This can become the difference between success or failure, and can determine whether an operation is a money-making opportunity. For example, some believe that United's and American's reservation systems not only provided specialized treatment to their agents, but also obtained considerable revenue (as high as 24 percent of the firm's profits) from access fees. Every aspect, from food ordering to flight coordination and personal scheduling, is under computer control. Their network's information volume has steadily increased.

Secure and survivable transport on higher and higher-speed networks was the objective for the '70s and '80s. Now, in the '90s and beyond, these systems need to be expanded to handle the airline industry's extensive mergers and acquisitions, as it settles on four or five big carriers whose fleet size and passenger volume must support the high debt status that they acquired during the turbulent '80s. (See FIG. 3-21.)

Tasks

The general functions for transportation industries are listed in TABLE 3-23.)

Table 3-23. Transportation Tasks

Transportation (air, rail, car)	Narrowband	Wideband	Broadband
Travel agencies inquiry	I/R-T-TS		V
Customer control	I/R-DC-DD	T-TS	
Operations control	I/R-DC-DD	T-TS	VC
Maintenance control	I/R-DC-DD	T-TS	VC
Logistics control	I/R-DC-DD	T-TS	
Personnel control	I/R-DC		
Billing/collection control	DC	T-TS	
Payroll/accounting	I/R-DC		
Administration	I/R-DC-DD	TS-RD	VC

Information users

Again, the key forms of data handling—I/R, DC, and DD, are prime candidates for executing narrowband operations over public/private overlay networks. Wideband transport will be needed between distributors, hubs, and maintenance centers with video broadband capabilities added later. (See TABLE 3-24.)

Table 3-24. Transportation
Narrowband-Wideband-Broadband Users

Narrowband	50%	I/R-DC-DD	Data users
Wideband	25%	Transitions	Data users
Wideband	20%	Time-sharing	Image users
Broadband	5%	Video conference	Video users

Information services

- Scheduling programs.
- Booking programs.
- Status programs.
- Maintenance programs, etc.

Information services—libraries, data centers, research institutes

Information providers, such as libraries, researchers, and data centers, will be providing their customers in the home or office with access to computer systems and databases. Later, these reports and summary analyses will be supplemented by image and video documentation. Information programs will, of course, need access to sophisticated computer systems via time-sharing and remote control. (See FIG. 3-22.)

Tasks

(See TABLE 3-25.)

Table 3-25. Information Services Tasks

	Narrowband	Wideband	Broadband
Library files for inquiry	I/R-DC		
Library-library info. exchange	DD		
Library-distributors info. exchange	DD-RD		
Library-distributors trans.	T		
Remote book review service	RD		V
Library-to-business terminal	I/R-RD		V
Library-to-home terminal	I/R-RD		V
Library micro dot book reading	RD		V
DP info. services (statistics)	I/R-DC	TS-G	V
DP walk-in/remote computer service	TS	TS-G	TS-G
Canned application programs	TS	TS-G	TS-G
Research remote computer-exchange	TS	TS-G	TS-G

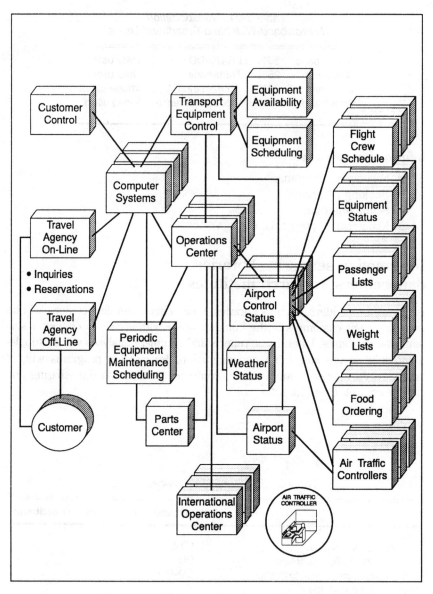

Fig. 3-21. Transportation (example airport).

Information users

Depending on whether the application is a library or a data center, information needs will vary from narrowband inquiry/response, data collection, data distri-

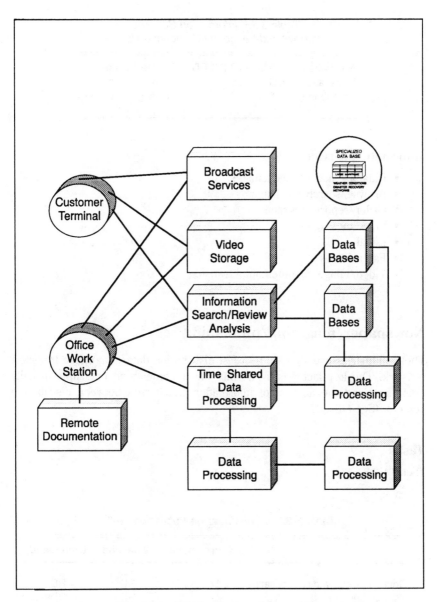

Fig. 3-22. Information service providers.

bution, and remote documentation (I/R, DC, DD, and RD) to more sophisticated high-speed wideband and broadband transport for time-sharing and
graphic operations with broadband visual capabilities for full document reading. (See TABLE 3-26.)

Table 3-26. Information Services
Narrowband-Wideband-Broadband Users

Narrowband	75%	I/R-DC-RD	Data/text users
Wideband	20%	TS	Graphic users
Broadband	5%	TS-V	Visual/graphic users

Information services

- Database access mechanisms.
- Browse/search mechanisms.
- Order/report programs.
- Voice messaging.
- E-mail.
- Fax facilities.
- Video/image documentation transfer.

Newspaper/magazine/publishing

These industries usually require a fast distribution data/text/image transport network. The key factor is availability due to time-frame pressures of daily operations and the amount of information transferred between distribution centers. (See FIG. 3-23.)

Tasks

(See TABLE 3-27.)

Table 3-27. News-Magazine-Publishing Tasks

	Narrowband	Wideband	Broadband
General news syndicate bulletins	DD	RD	RD
Remote printer	RD	RD	RD
Customer bills	DC		
Distribution control	DC-DD	DD	DD
Advertising	DC-DD	RD	RD
Reporter-staff info. exchange	DC-DD		
Research staff operations	I/R-DC-DD		

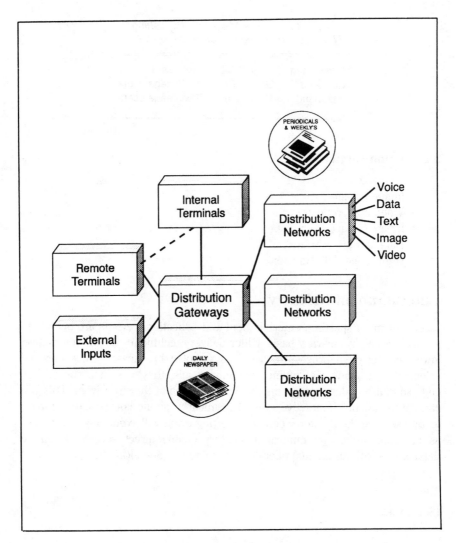

Fig. 3-23. News/publications.

Information users

While information is being collected for insertion, universal-global narrowband networks will comfortably support these needs as long as quality is achieved. However, at the time of distribution of large blocks of information for remote printing, there's a need for high-speed, high-quality transport facilities between centers. (See TABLE 3-28.)

Table 3-28. News-Magazine-Publishing Narrowband-Wideband-Broadband Users

Narrowband	70%	DC	Text users
Wideband	20%	RD	Text/image users
Broadband	10%	RD	Text/image users

Information services

- Text transfer.
- Image transfer.
- Video conferencing.
- Image storage/retrieval.
- Data storage/retrieval.
- High-speed data transfer.

Entertainment industry

More and more sporting, entertainment and educational events are becoming available on a pay-per-view basis. Older delivery techniques use an interactive, low-speed selection process to access scrambled broadcasts. Once viewers' selections are made via "offline" telephone calls, they're sent a special key code so that broadcast programs can be decrypted at their receivers. This will change to "online" interactive systems. Similarly, phone conversations for selecting seats at the theater, opera, or sporting event will eventually go "online" as personal home communication centers enable direct access to remote databases—without manual operator intervention. (See FIG. 3-24.)

Tasks

(See TABLE 3-29.)

Table 3-29. Entertainment Tasks

	Narrowband	Wideband	Broadband
Sports setting	I/R-DC-T		
Theater setting	I/D-DC-T		
Performance bookings	I/R-DC-T		
Sporting events—home*	I/R		RD/V
Entertainment events—home*	I/R		RD/V
Education events—home*	I/R		RD/V

*Broadcast or on demand

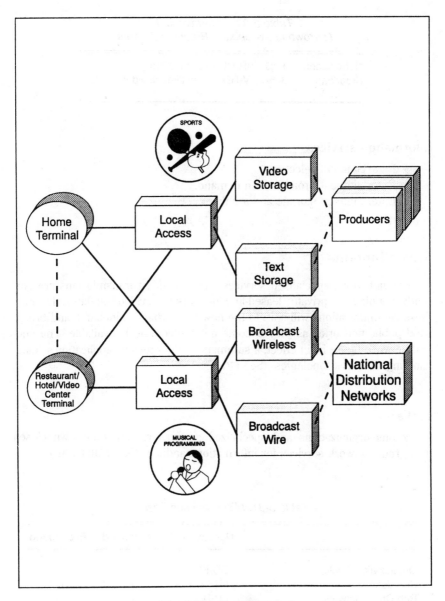

Fig. 3-24. Entertainment.

Information users

While the interactive booking and selection process is really handled by narrowband networks, the actual viewing of the event requires broadband—with better and better resolution. Information is sent in the broadcast mode. Later, when more channels will be available, this will enable individual customer selection in the "on-demand" mode. (See TABLE 3-30.)

Table 3-30. Entertainment Narrowband-Wideband-Broadband Users

Narrowband	10%	IR-DC	Data users
Broadband	90%	V-RD	Text/image/video users

Information services

- Entertainment selection.
- Entertainment broadcast/on demand.
- Entertainment storage and archiving.

Miscellaneous

Overlay networks, specialized services, and broadcast transmissions are currently handled by private, leased-line networks or video satellites. In time, these communications will shift to the new narrowband, wideband, and broadband public networks, as they become more ubiquitously available. The general network figure is, in effect, a summary of the previous application's "data-handling" network topologies. (See FIG. 3-25.)

Tasks

Numerous organizations and specialized services require their own closed user-group network services for information-handling. (See TABLE 3-31.)

Table 3-31. Miscellaneous Tasks

	Narrowband	Wideband	Broadband
Organizations' bulletins	DD-EV		
Civil defense	I/R-DC-DD	T	VC
Red Cross network	I/R-DC-DD	T	VC
TV transmissions			V/VC
Political organization nets.	I/R-DC-DD		VC
Government services	I/R-DC-DD	T	VC
Postal messages special delivery	DD-RD		
First class business messages	DD-RD		
First class home mail	DD-RD		
Computer-to-computer data banks		DD	

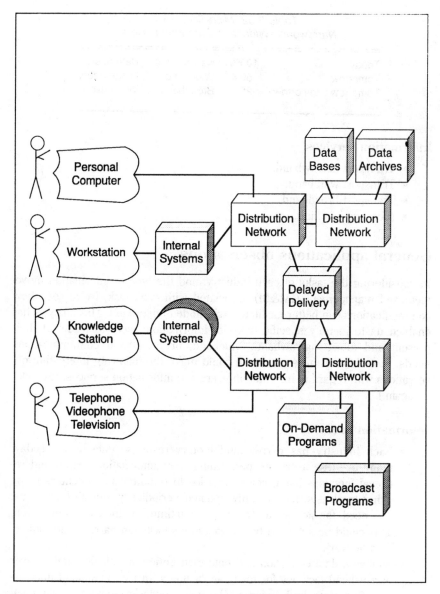

Fig. 3-25. General network.

Information users

Information users will grow from narrowband data to wideband image, to broadband graphic, to still frame, and then to full-motion video as these networks become more and more economical and ubiquitously available. (See TABLE 3-32.)

Table 3-32. Miscellaneous Narrowband-Wideband-Broadband Users

Today	100%	Narrowband	Data users
Tomorrow	50%	Wideband	Image users
Tomorrow's tomorrow	25%	Broadband	Video users

Information services

- Transport narrowband.
- Transport wideband.
- Transport broadband.
- Message handling.

General applications observations

In considering the tasks of each industry, and the type of information movement and management (IM&M) associated with each task, we've seen how communications can better facilitate each mode of operation. This analysis has enabled us to identify specific types of information users in terms of their present and future narrowband, wideband, or broadband communications needs. In reviewing these users, we could make the following interesting observations with regard to information users and information services. (See FIG. 3-5 again.)

Information users

- Each industry has narrowband inquiry/response data users. Today, many of these users are performing their information search and retrieval functions using voice inquiries to remote service centers or by using local lists that are only updated periodically; but all have noted the need for access to "online," "real-time,""up-to-date" information. This could be achieved by an ubiquitous switched narrowband, public-data network.
- Similarly, data collection and data distribution are nicely handled over narrowband facilities for many applications. In fact, models of information flow over high-speed SMDS-type networks show the additional overhead added to the exchange of these packets of information. This indicates an opportunity to transfer these types of information over narrowband or wideband data circuit or packet-switched facilities.
- Time-sharing and remote documentation can be supported by both narrowband and wideband facilities, depending on the amount and type.
- Time-sharing operations will be the natural, national extension of the computer into every industry and business. Thus, the volume of these calls will be extensive. The data transfer operation can be reduced by

more sophisticated terminals, which will increase transmission data rates and reduce calls to the central computer. Here, the user could be dropped from the network during idle time, but that will require fast circuit connections or fast packet handling.

If the bandwidth is increased, large blocks of remote access data will move more quickly through the network. So, this will be a candidate for wideband networks, but the data will be in bursty, variable-bit-rate form.

A switching system with normal connect time can serve both remote access and time-sharing users, until it's replaced by a fast connect/fast packet switching system, as usage volume increases. However, if network costs for time-sharing don't become competitive, as time-sharing becomes more conversational, this service will remain on private networks.

- Videophone, video conferencing, high-resolution graphics, full-motion/ high-resolution education, entertainment, and sporting events will require greater bandwidth facilities. Broadband usage will shift from point-to-point or point-to-multipoint LAN-type networks to fully switched public broadband networks, which provide for ubiquitous high-volume, high-growth transport usage.

- As success begets usage, which begets more usage and more success, higher-speed transport facilities need to be available to transport narrowband and wideband information in order to ensure that there are no delays, blockages, and queuing.

- In terms of percentages of information users for each of the three types of networks, it's interesting to note that initially, 70 percent (or 7 out of 10) of today's users can be satisfied by a robust, ubiquitously available public data switched narrowband network. Without requiring a picturephone in every home, the next major grouping of applications can be appropriately handled by wideband facilities. LAN interconnection needs can change, as internal CPE systems begin to switch data, and especially as they also shift to switching variable numbers of channels over the wideband rates. Hence, there will be a need for public wideband switching facilities. In time, there will be several shifts in narrowband users so that eventually, approximately 5 out of 10 may be switched narrowband, 4 out of 10 switched wideband, and 1 out of 10 broadband.

- The alternative to the narrowband/wideband services will be fully deployed video through fiber-to-the-home methodologies—if indeed we have a fully "fibered" country by 2015. If so, then by the turn of the century, 3 out of 10 may be provided fully switched public broadband capabilities, rather than the preceding 1 out of 10 estimate, which was for applications over selective, more privately tailored facilities. Then, in time, perhaps by 2031, 7 out of 10 users will be on broadband networks.

- "Broadcast broadband users" are natural for point-to-multipoint deployment in the form of cable TV systems, VHF/UHF, and satellite systems.

However, "interactive broadband users" are natural for switched fiber. Once the fiber is deployed, it can handle broadcast needs either on an overlay basis, as fiber capacity is increased, or on a switched basis as interactive selection and browsing mechanisms are provided using both lower- and higher-speed transport capabilities.

- Attempts on the network (set ups and take downs) and throughput capacity are the two key factors that affect network availability. Here, balance is needed in network architecture to encourage growth and usage, but we must ensure network integrity and survivability as information usage increases with more and more users. The increased usage will affect connect time, holding time, error rate tolerance, terminal device speeds, operation mode, and facility routing and addressing requirements, as noted in the appendix.

Information services

Voice, data, text, image, and video services for today and tomorrow have been indicated for several applications here, as well as in the appendix. These services include: voice messaging, CLASS, E-Mail, fax, alarm monitoring, environmental control, meter reading, and message transport (broadcast, polled, and randomly generated). These will be provided over narrowband facilities that will enable access to advanced information service nodes for further services such as: encryption, store and forward, delayed delivery, gateway access to databases, protocol conversion, code conversion, presentation and data manipulation mechanisms.

LAN-to-LAN wideband services now use for frame relay, SMDS, and FDDI transport and switched wideband facilities. These will provide customer access and control of virtual private network configurations for image, X-ray, graphic, computer art, and computer-to-computer time-sharing and distributed processing.

Videophone, video conference, video mail, 3D imaging, full-motion, high-definition entertainment, education, and sporting events require extensive bandwidth and sophisticated advanced service nodes for storing, accessing, manipulating, and presenting these complex forms of information (see the appendix, Information Services). (See FIG. 3-26).

Conclusion—The information marketplace

Number please?
Information please?

When telecommunications companies built a voice network for, "Number please?," customers first talked to the operator to be connected to a neighbor. Later, companies used advancing technology to shift to automated switching of calling and called parties. Today, when customers call a ticket operator for

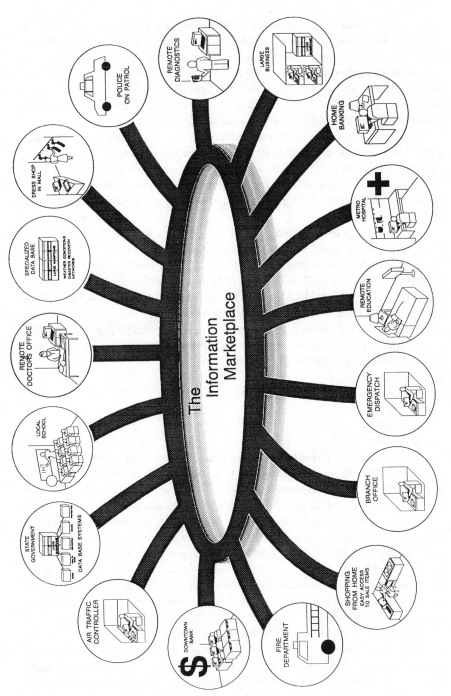

Fig. 3-26. The Information Society.

"Information please?" seat selection and reservations, they'll begin to shift to "online" information switching, thereby enabling anyone in any place to obtain direct access to remote databases such as theater seating. So, we've entered a new arena in a new marketplace. It's analogous to the new rules of bridge, as bridge players today shift from the old Goren methods to new American Standards rules. Today's players know it's a new game when we say, "Bridge anyone? Anyone for bridge? Do you want to play a game?"

Similarly, the information marketplace is like a new game. Every new game has its set of basic rules that are used when the game is played. For example, bridge is a game with various bidding conventions. As players are dealt different hands, the rules and strategies are drawn upon and used, depending upon the hand dealt and the skill of the player—as seen in the playing of the hand.

So, in viewing the new information marketplace, we have basic technologies—narrowband, wideband and broadband, which are similar to the suits—clubs, diamonds, hearts, and spades. Each set of face cards, in this analogy, corresponds to n-ISDN, W-ISDN and B-ISDN capabilities. It's now up to the marketing player to establish the best solution for the different market sectors and industries. Each of these "hands" changes over time and requires different solutions. The information marketplace is a new and changing game. To play it, we need to learn its rules, conventions, suits (technologies), and strategies, as we develop new skills for playing the game.

Playing the game
is indeed very complicated,
taking time,
knowledge,
and patience.

Similarly, as we move from the telephone (voice only) to the dataphone, to the videophone, we need to understand the specific requirements for the various types of data, text, image, and video multimedia applications. This is a new industry; it's time to use today's technology to meet immediate needs to encourage user acceptance and growth. It's time to add feature on feature, with a twist here and a turn there, to tailor network offerings to their specific applications. It's a time for suppliers, providers, and users to work together to identify the technology that they'll be needing in the future. It's time to establish a realistic family of equipment specifications for the switching, transport, and CPE systems that providers will need in order to meet their customers' needs in a timely fashion.

To add form and substance to the shapeless monolith, it's time to visualize market opportunities in terms of an expanding array of information services. Beginning with various voice enhancements, it is possible to grow through data, text, and images to global video offerings (see the appendix—Information Services). Figure 3-27 shows where new services blossom to meet the somewhat singular but cross-related applications of the major industries, ad-

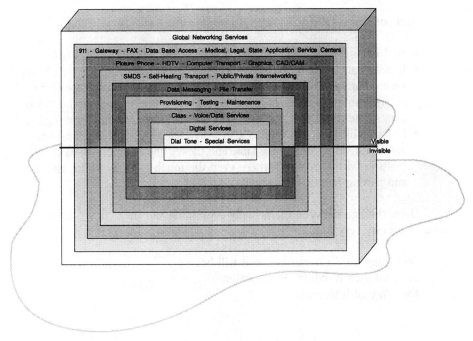

Fig. 3-27. Information services.

dressing the personal and operational aspects of our human needs and aspirations.

Thus, the future millennium may see a society in which we achieve such capabilities as visualized by Alan Chynoweth, who, as Bellcore's vice president of Applied Research endeavors, is responsible for exploring future unknown technical possibilities:

- Personal communications via personal number calling.
- Distributed workplace via sophisticated workstations and personal computers.
- Multimedia services for the "home of the future/office of the future" via "fiber to the home/fiber to the desk" technology, enabling HDTV, artificial reality, voice sensitive computer responses, etc.
- Medical advances in the conquest of diseases via sophisticated imagery and computer analysis.
- Increased food supply via genetic engineering.
- Alternative energy sources using synthetic fuels and solar energy.
- Educational advancements via computer aided instruction and personal development tools.

And perhaps someday, as others have noted:

- Hydrogen-powered automobiles, electric cars, and high-speed trains.
- Usable, safe, reliable, stable nuclear energy sources without dangerous waste. For example, when fusion or large-scale commercial breeder reactors can be successfully achieved in secure plutonium recycling plants, and these new nuclear parks supply power for an entire region.
- Holographic video conferences, in which each participant sits around a simulated video conference table and moves and functions in extremely realistic "virtual reality," where even the touch and feel of the opening and parting handshakes would seem real.

Here, the line between the actually achievable and the actually unachievable is actually very thin . . .

So it may be, so it can be, so it will be,
for centuries to come, as we enter
the "Age of Information" . . .

4

Information networks

"We had better establish where we are going,
so we will know that we have gotten there,
when we have arrived."

Dr. Doolittle

Introduction

We've moved from customer needs to applications to possible new services
that meet these needs. The appendix A tables provide lists and lists of possibilities, opportunities, and the new network and service platforms to achieve
the possibilities. So, what networks should be established? What steps should
we take to restructure America's telecommunications infrastructure?

We've seen the needs and we've identified potential applications. Once we
know where we're going and the risk is understood and somewhat minimized,
we should be more willing to take the first steps to get there. So, where are we
going? What information networks are needed? (See FIG. 4-1.)

When we speak of data and what's happening with regard to its impact on
the voice-based telecommunications industry, this brings to mind the story of
two neighbors who built beautiful houses looking out over the ocean. One day,
the owner in one home, which was more privately located high above the
other, asked his neighbor to walk around his yards to see if she noticed something different. She did; the grass felt a little spongy; it seemed to be giving a
little more than at the lower place, which was more publicly located on the flat,
dry, open land. In time, the higher land began having different types of flowers
and plants. This was attributed to the excessive amount of mountain water that
was contained in its mountain lakes.

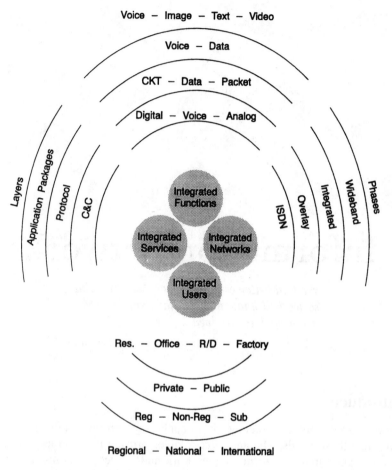

Voice – Image – Text – Video

Voice – Data

CKT – Data – Packet

Digital – Voice – Analog

Layers

Application Packages

Protocol

C&C

Integrated Functions

Integrated Services

Integrated Networks

Integrated Users

ISDN

Overlay

Integrated

Wideband

Phases

Res. – Office – R/D – Factory

Private – Public

Reg – Non-Reg – Sub

Regional – National – International

Fig. 4-1. Network services.

Eventually, one night after a severe storm, the whole front yard slid off the side of the mountain in a wall of mud. As it tore away, it buried the yards and home of the neighbor below it. For over the years, the unchanneled, dammed water had permeated, saturated, and eventually, totally undermined the entire area . . . So it is with data, as it seeps from its separate, aloof, private world into "Ma Bell's," GTE's, and the independents' common carrier public voice networks. Properly managed and channeled, it could be a great blessing and a great opportunity. Ignored and undirected, it can produce devastating results on a traditionally stable voice environment, perhaps even causing the loss of one's previously beautiful domain. No area is safe, as the analogy notes, even the higher, private home is now slipping off its high, separate perch to the public lands below . . .

So, even if you think it's not a concern in your own back yard, you might be surprised to wake up one day and find that it did affect us. Your own yard or home could be gone. With these thoughts in mind, let's ask, "What is data? What are our data strategies? What should our data strategies be?

Data strategies—information strategies

As noted earlier, data originally could be viewed as information that's created, stored, accessed, manipulated, and processed to provide mathematical and statistical analyses. However, over time, data has come to mean much more than that. The role of data in numerical analysis has expanded to include every aspect of the information marketplace. Here, we entered the world of the transfer, storage, and word processing of various forms of textual information. This gave way to imaging, graphical displays, and computer art. Then there are video variations such as still frame, freeze frame, low-resolution video, or full-motion, high-resolution communication. Here, the telephone gives way to eye-to-eye videophone.

The original functions of data as statistical and numerical information in databases were really the stepping stones to numerous forms of information that's now all-encompassing, including voice, data, text, image, graphics and video entities. So, today, data handling can be viewed as "information handling."

Hence, traditional "data strategies" for public or private networking are really a subset of "information strategies," where, as these boundaries blur, "information handling" can be initially viewed in terms of the current and more precise and realistic transport limitations of today's media. This will give way to the less-clear, less-restricted, less-bound possibilities of tomorrow's media. Data then begins with its numerical analysis applications, but expands to include all forms of information handling as narrowband, wideband, and broadband information services opportunities are pursued.

Therefore, we need to maximize today's voice network capabilities, by expanding it to include, as much as possible, the movement and manipulation of data. This can be considered the first of several forms of total "information handling" using the narrowband information transport mechanisms. Here, we need a data-handling transport strategy that provides for the public movement of data, and that facilitates its transport, processing, manipulation, and presentation to the new information users.

This will be a *public data switched network*, with all its store and forward, error correction, delay delivery, addressing, and alternate routing capabilities.

This strategy will then be expanded to encompass the movement of all information, which will require greater capabilities than the current public transport plant. These opportunities are divided into two parts: wideband and broadband. Wideband is the bridge between the old and the new. It will span the traditional data-handling capabilities of simply overlaying "black boxes" for additional data message switching. Special conditioning of the copper network will be added in order to reach its maximum speeds (of 1.54 to 6.3 Mb/s).

Later it will be expanded to include the first offerings of switched fiber at T3 (45 Mb/s). Thus, we've moved into the domain of wideband "information-handling" services. These are mainly transport flexibility and selectability. Here, the customers can interconnect their local private networks, called LANs, together over shared public facilities, to changeable destinations, selecting as much transport capacity as desired, and paying only for when and where they use the services.

This selectability then extends, into higher and higher capabilities, to the movement and manipulation of broadband information. New SONET-based bandwidth will enable a new host of broadband capabilities, which will subsequently encompass earlier narrowband and wideband offerings, as virtually unlimited transport (in the order of 155 Mb/s and 600 Mb/s) is available to the customer. Thereby, for example, enabling full-motion video. Hence, our data strategies are really part of a three-part "information-handling strategy"— narrowband, wideband, and broadband.

Narrowband

This will include the Public Voice Network, expanded to include CLASS, voice messaging, fax, etc., and the Public Data Network, using traditional copper plant with narrowband ISDN transport and switching to enable circuit and packet switching for message switching, E-mail, broadcast, polling, alarm monitoring, energy management, meter reading, etc.

Wideband

This will include the Public Wideband Network to enable the customer to actively select bandwidth over variable routes in the T1 and T3 range on a selective location basis.

Broadband

Here, information can be in full-motion, high-resolution graphic/video form. Transport will use the SONET 50+ Mb/s multiples and B-ISDN standard interfaces of 155 and 600 Mb/s to enable selective customers access to high-speed offerings for computer-to-computer communication, videophone, graphic workstations, and high-definition, high-resolution television applications.

As a parallel strategy, private-to-private wideband T1/T3 internetworking needs can be met, initially using frame relay and SMDS transport techniques. Later, these offerings will expand to include full broadband capabilities on a selective basis.

As an enhanced information services strategy, service nodes, using new information switches, will facilitate application service centers that provide enhanced data-handling/information handling features and services such as fax, store and forward, data encryption, protocol conversion, delayed delivery, image files, and gateway access to specialized databases. (See FIG. 4-2)

So, lets take a closer look at what's needed by the turn of the century.

Network of networks

In considering the layers of services indicated by the market assessment of customer applications requiring voice, data, text, image, graphics, and video

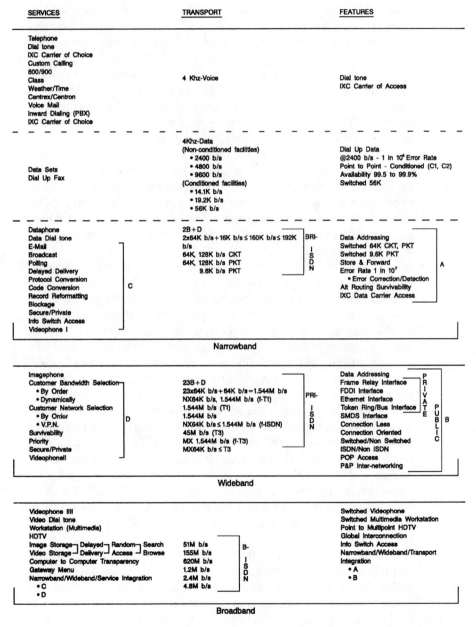

Fig. 4-2. Information networks' services.

multimedia communications, it's time to identify the supporting network infrastructure.

In reviewing FIG. 4-2, it's quite apparent that we need:

- An expanded voice/audio network infrastructure that enables new, versatile voice message services to be overlaid on existing capabilities to form an extended and enhanced Public Voice Network for the '90s. This will maximize our existing narrowband network voice offerings by using new ISDN (access), SS7 (signaling) and IDN (Integrated Digital Network) technologies to upgrade the current infrastructure to provide expanded offerings, such as Voice Mail, calling party ID, CLASS services, selected call (transfer, blockage, message, routing, and priority override), Intelligent Network 800, 900 database access services, special audiotex services, fast provisioning, number change, etc.
- An infrastructure that fosters the economical movement of low- to medium-speed information over our current copper facilities to formulate the Public Data Network for the '90s. This will establish a new family of narrowband data services over the existing copper facilities, using ISDN data interfaces to enable both the circuit- and packet-switched movement of data from terminal to computer, computer to computer, and terminal to terminal.
- An infrastructure that enables the internetworking of private facilities over unlimited, high-speed, secure and survivable (S&S) facilities on a demand-type usage basis to form the Public Wideband/Broadband Access Transport Network for the '90s. This will provide an exciting group of wideband services ranging from wide area networks that provide a variable number of 64K b/s channels up to T1/T3 rates of 1.5/45M b/s for data, imaging and graphic display of information. These will be used for applications such as: X-rays for the medical industry, stock forecasting for the securities industry, CAD/CAM for the manufacturing groups, WANs that internetwork LANs and foster the movement of information between private networks and public networks by providing new, high-speed data transport services such as SMDS, Dynamic Bandwidth, Bandwidth on Demand, Survivable Transport, Improved Error Rates, Access to Multiple National Carriers and International Value Added Networks. In time, these platforms will evolve to full B-ISDN/SONET/SDH broadband switching nodes using self-healing, survivable fiber deployment topologies that locate these new switching centers closer to the customer.
- A revolutionary network infrastructure to encourage the growth of many new high-information-content services that require substantially higher bandwidth, faster movement, and lower error rates than obtained by evolutionary changes to the current network (as it becomes a new Public Broadband Network by the turn of the century). This will achieve broadband services such as: High-Resolution Video Services

that enable visual imaging, desktop conferencing, high-quality Graphic Display, Computer-to-Computer Data Exchange, Video Displays for Education, Media Events, Image Archives, and Entertainment-Type Video Services. These will be offered over a revolutionary new fiber-based voice, data, image, video switched network with automated support capabilities that establish extended/enhanced services in the nonregulated community via an overlay of Information Services such as Info Switches that provide direct access to Databases and Advanced Data-Handling Services or, via Gateway Access to Stand-Alone Application Service Centers. These will provide special services such as E-Mail and access to CPE service centers for specialized/unique databased services such as, poison control centers, patient records, etc.

Network architecture

To achieve these new narrowband, wideband, and broadband voice, data, and video network services, the public network will require a new architecture. This will be especially apparent as changes to more functionally facilitate private and public internetworking take place. These changes must continue to take into account shifting regulatory/nonregulatory issues and boundaries as the RBOCs are and are not allowed to provide various information "handling" services. Another major change is occurring in the leased-line/trunk-type special services. More and more users are requesting control over their network configurations. These services are being automated to enable dynamic setup and drops in order to achieve virtual private networks (VPN) to here and there as needed. Further, architectural changes are required to enable more or less capacity (transport) to be available, as needed, so that not only new routes are dynamically established, but also varying amounts of capacity can be requested by the user before transfer or even during transfer.

Concern for bottlenecks, survivability of central office fires, the need to quickly access an IXC's POP, a request to be routed directly to an advanced services platform, and the need to reduce unshared transport distance to reduce access cost all lead to the need for a new network architecture. Here, the traditional five-level Class (1 to 5) central office structure, which was based on voltage net loss (VNL) distance requirements for transporting analog signals, must soon give way to new switching systems located closer to the customer. Fiber can now be deployed in survivable rings with access nodes for private-to-public networking. These nodes need to become network-level switches that can provide initial address translations in order to route calls (voice, data, or video) to the appropriate network service provider.

Hence, the appendix provides an in-depth look at new network products such as Class 5s, Class 6s, Class 7s, and information service nodal switches (Info Switches) that provide the building blocks for the future network architecture (See appendix—Private and Public Networking). See FIGS. 4-3A, 4-3B, and 4-3C.)

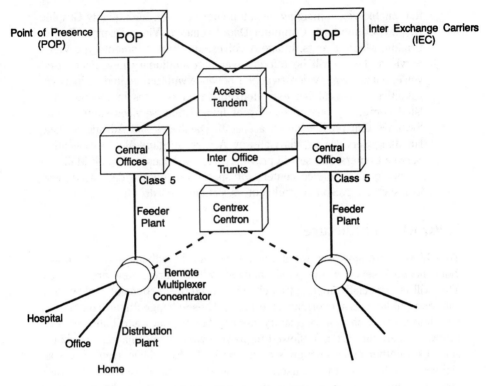

Fig. 4-3A. Telephone company today.

Network products

As we develop new products to overlay wideband and broadband services, it's important to differentiate functions and time frames for these new switching entities. The information-services switching platforms will provide services for all three transport areas. In this manner, there can be a separate info switch that has an associated application service center providing shared databases and third-party software programs. The info switch would be specifically tailored for the narrowband services of the public data switched network. There can also be an info switch that enables wideband, specialized private internetworking functions, as well as one for video-based services such as video messages.

While the network is being deployed in the narrowband arena, the new wideband (later broadband) access switches (Class 6) can be established in the urban community on overlay facilities. They will later function as front-end nodes, for future Super5s (Class 5s), when these new broadband switches begin supplying ubiquitous videophone-type services. At the same time, customer premise systems (CPSs) will evolve through the various transport bandwidth configurations, as new network interface standards are employed to en-

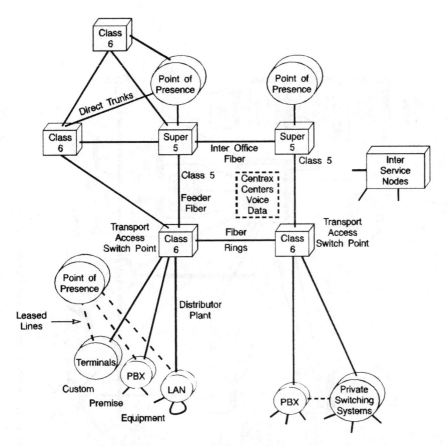

Fig. 4-3B. Telephone company tomorrow. (Public Basic Transport Company, PBTC.)

able a full range of varying bandwidth switched customer services (Class 7s). In turn, the rural community (specifically the county seats) can be overlaid with new Class 6 switches (as can numerous central offices in the large urban communities) as they cap their digital remote switch unit-base configurations for narrowband services and overlay wideband-broadband rings that home on new Super5s centered somewhere in the rural region.

In this manner, both rural and urban communities achieve the public data, wideband, broadband transport infrastructure identified in previous discussions. Here, the maximum growth of new offerings is achieved from the existing copper plant. Meanwhile, new fiber is slowly and selectively distributed into the residential marketplace as the industry attempts to achieve the 2015 goal of full fiber deployment to the home.

Support systems must also track these networks to enable full automation of provisioning, testing, maintenance, and administrative functions. This will require considerable open network architecture (ONA) work to enable infor-

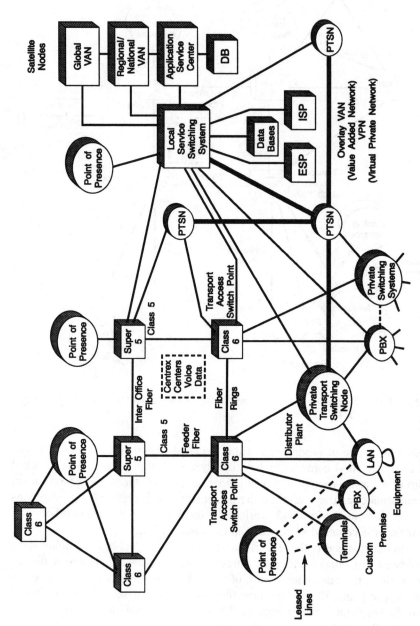

Fig. 4-3C. Future information switched services company.

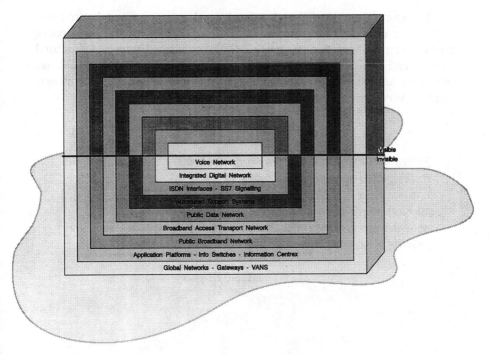

Fig. 4-4. *Network of networks.*

mation networking architecture (INA) interface standards to establish traveling call status requirements to direct further distributed service offerings to and from the various information service providers (ISPs) and enhanced service providers (ESPs). (See ONA analysis in appendix A.)

In this manner, the new network architecture, which will provide for growing transport capacity and advanced services under customer control, will be achieved by new CPE/CPS switches (Class 7s), access switch nodes (Class 6s), supercentral office switches (Class 5s), information switching platforms (info switches), and application services centers (ASCs).

Conclusion

In the "information technologies" section of this book, we "tickled our minds" with the narrowband, wideband, and broadband technical possibilities of these networks. They offer a host of new features from which the desired information services can be derived. (See FIG. 4-2 again.) Packet, fast packet-Asynchronous Transfer Mode (ATM), circuit-Synchronous Transfer Mode (STM), and variable channel switching possibilities using silicon and optical techniques need to be resolved over the '90s for the various bursty, variable-bit-rate (VBR) and continuous-bit-rate (CBR) services. (See FIG. 4-4.)

Therefore, these market and network needs require a new layered networks' layered services infrastructure solution that supplies narrowband services over existing facilities that are overlaid with conditioned transport wideband services and then augmented by a growing array of fiber-based, broadband services to formulate an expanding and encompassing broadband network services infrastructure.

"Telecommunications will play a critical role in the global economy, while developing nations need a basic network to provide the foundation for growth, developed nations need a sophisticated infrastructure to provide the competitive edge."

Carol Wilson
Editor, *Telephony*

Part III

The emerging
global
information society

A time of change,
A time of complexity,
A time of opportunity . . .

5

The global information society

"It was the best of times,
It was the worst of times,
It was the age of wisdom,
It was the age of foolishness,
It was the epoch of belief,
It was the epoch of incredulity,
It was the season of life,
It was the season of darkness,
It was the spring of hope,
It was the winter of darkness,
We had everything before us,
We had nothing before us,
The period was so far like the present . . .
It was a time of great expectations . . .
. . . filled with uncertainty . . ."

Charles Dickens
1859

Society continues to advance in both time and space. With the passage of time, remarkable advancements for a more civilized culture are slowly, but steadily, spreading to diverse geographical regions. These advancements take the form of education, employment, housing, food, medicine, personal wealth, goods, and entertainment. Over the last few decades, the Japanese have achieved economic success as they've launched an extensive array of new, applied research and applied development programs with emphasis on quality.

This severely challenged and defeated the '60s-to-'80s U.S. production mentality of "No need for new products!," "Ship it and fix it in the field!" or

"Cost reduce it; who cares how long it lasts?" So, the game became serious in the '90s as manufacturing jobs severely decreased and the wobbling trade imbalance became an impossible balancing act. As people become more and more disgruntled with the realities of eking out a living on service industry salaries (many at half the wage of previous manufacturing jobs), the president and Congress must face the grim reality that U.S. industry is in the process of "giving away the store."

However, over the late '70s and throughout the '80s, several major, shifting factors forever changed the role of large industrial nations in the international game. The first was the massive amount of technology transfer. Multinational firms, such as ITT, hired the best that U.S. design firms had to offer. They became expatriates working abroad, in their foreign units, where they developed similar versions of products that they had designed earlier in their respective American firms. These foreign units, through natural localization practices, transferred this technology to numerous internal programs—thus capturing the transferred technology. Also, American graduate schools train many foreign students in the sciences. Similarly, numerous engineers from many countries attend scientific conferences, review trade journals and analyze patents as technology is interchanged and transferred around the globe by computer and communications technology.

The second major shift to take place is in the use of information to achieve greater market penetration with more cost-effective products. As more and more firms obtain the technology base needed to expand, enhance, and compete, the marketplace has become quite competitive. Today, information is used to automate assembly lines and control machines to such a high level of precision that older tool-and-die processes can no longer compete. Information is now used to determine exactly what the buyer wants. This is so perfected that production-line products are specially equipped with extensive ergonomic interfaces and operations that fit the users' exact needs—to the point of customizing each product for its particular user.

Hence, technology transfer and information use have become the driving forces for a new world marketplace. In some countries, workers were initially trained to assemble the new technologies for other countries, but as soon as local markets developed the countries became their own best users. Third-world countries with limited resources saw the less capital-intensive information market as an attractive alternative to complex, highly expensive industrial manufacturing lines. They found competitive advantages in using their inexpensive labor market to perform data input and formatting functions. Others, such as Singapore and Hong Kong, pursued enhanced financial services. Still others selectively established highly specialized product lines to become the "best" producer of this or that particular product in the world. And others recognized the economic advantages of providing communication gateways, as global satellites enabled economical, direct wireless access to their service nodes.

World spending for telecommunications reached new heights of $45,336 for 1991 and $46,100 million for 1992. As privatization, the third factor, traveled around the world to countries such as the United States, England, Japan, Singapore, Western Europe PTTs, South America, and New Zealand, the competitive market forces reestablished focus on customers with private ownership of the communication networks to better meet the customers' high expectations.

Another key shift came in the computer transformation from single, large, expensive mainframes to small, powerful, personal computers that are interconnected to share complex problems and information. This process, as noted in chapter three, has found its way into every task, every operation, every application of every industry. These distributed computers have an insoluble taste for more and more information, enabling more and more technology transfer and information usage. With access to information comes new knowledge and more unrest. With this power comes the break up of totalitarian-style dictatorships toward more democratic processes, which encourage the use and dissemination of more and more information. The result is a closer integration of countries throughout the world. Indeed, this becomes the final factor that will enable society to progress to new heights and achievements.

Hence, the stage is set for the economic wars generated from the new market forces unleashed by the preceding shifts. Together, they all become a huge wheel, getting bigger and bigger, gaining more and more momentum as each entity feeds on or builds upon the other to create more and more pressure to encourage more and more effort to achieve a more competitive, successful position in the expanding information-based marketplace.

Shifting markets—phasing networks

When discussing leapfrogging technology, it's important to appreciate the analogy of the needing to walk before running. There's a reason for the different and phased narrowband, wideband, and broadband deployment of the various technical functionalities. With narrowband comes a public data network operating at speeds of 64K and 128K b/s. These speeds will more than satisfy initial, small amounts of data movement. The time and location of availability become tied to economic justification, instead of technical feasibility issues, as we consider the ramifications of ubiquitously deploying a worldwide copper-based public data network or a worldwide, ubiquitous fiber video network. Countries may be more and more reluctant to foster telecommunication solutions over other internal, competing infrastructure needs such as roads, water, sewers, and food distribution.

We must be careful of tulip-mania or info-mania. Once, in Holland, a particular tulip became so popular that everyone decided to grow it. The growers believed that they would make substantially more profit the next year than with the other bulbs. Unfortunately, everyone grew the same type of bulb, and unfortunately, the world didn't appreciate it as much as the Dutch. The result was overnight disaster, as the marketplace reflected their overestimation. The severe correction, caused considerable losses.

That's not to say that an information explosion is not about to take place; it is. But, not everywhere, at the same time, in the same way. It will take time to establish a public data network (some in the most hostile of environments) around the world. It will take time to change from pricing on bandwidth to pricing on services in order for video service to become economically feasible. Unfortunately, or fortunately, depending on one's perspective, this will further encourage customers to use private overlay networks and wireless satellite networks. (See FIG. 5-1.)

Hence, the information revolution is indeed happening, but the secret to success is in determining where and when it's happening. It will not happen with the same degree of activity everywhere. Therefore, it's extremely important to determine the following: What countries will recognize the power of economically priced services in order to obtain higher demand for high-bandwidth video, resulting in a more ubiquitous broadband infrastructure? What countries will encourage the deployment of a public data network? What countries will foster such an open structure that nothing works? What countries will only foster private solutions for large business? What country will . . . ? Indeed, these are some of the basic questions identifying where and when the Global Information Society will develop.

On the other hand, it will indeed be exciting to see it expand and develop as it twists here and turns there along its path to maturity, influenced by economic necessities, increasing populations, and increasing national pressures to compete with successful products. So, the Global Information Society will blossom, sometime in the next millennium, the third millennium, the Information Millennium.

"In another moment
Alice was through the glass,
and had jumped lightly down
into the Looking-Glass room . . ."

Lewis Carroll

So, let's close with a look forward at The Age of Information . . .

It is 7:00 a.m. in Vancouver and Seattle, 8:00 a.m. in Edmonton and Denver, 9:00 a.m. in Minneapolis and Chicago, 10 a.m. in Toronto and Washington, 3:00 p.m. in London, and 4:00 p.m. in Geneva. International Information Gateways (IIG) has just initiated a videophone conference call to discuss an exciting new venture. It is not yet the typical universal communications form of the early 21st century, but IIG specializes in advanced information movement.

After the turbulent years of the '90s, one of the U.S. regional holding firms made two strategic decisions. It was going to create a local information society in a series of newly planned international cities, near each of its major cities, by providing within them extensive information-handling network services.

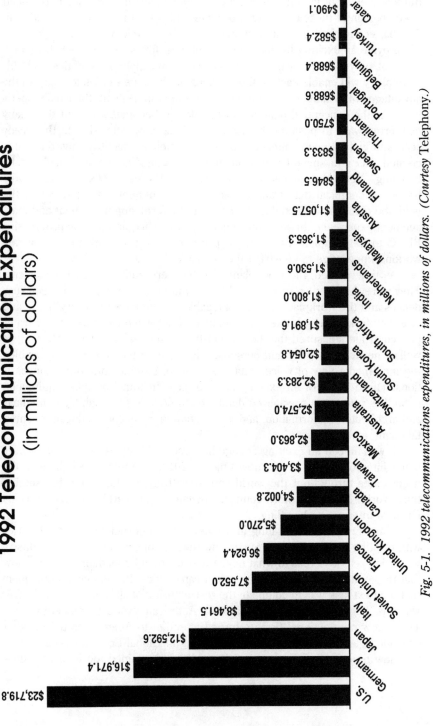

1992 Telecommunication Expenditures
(in millions of dollars)

- U.S. $23,719.8
- Germany $16,971.4
- Japan $12,592.6
- Italy $8,461.5
- Soviet Union $7,552.0
- France $6,624.4
- United Kingdom $5,270.0
- Canada $4,002.8
- Taiwan $3,404.3
- Mexico $2,863.0
- Australia $2,574.0
- Switzerland $2,283.3
- South Korea $2,054.8
- South Africa $1,891.6
- India $1,800.0
- Netherlands $1,530.6
- Malaysia $1,365.3
- Austria $1,057.5
- Finland $846.5
- Sweden $833.3
- Thailand $750.0
- Portugal $688.6
- Belgium $688.4
- Turkey $582.4
- Qatar $490.1

Fig. 5-1. 1992 telecommunications expenditures, in millions of dollars. (Courtesy Telephony.)

But a key element in constructing a competitive local marketplace is the need to equip each of these internetworked, new cities with instantaneous, full multimedia, economical communication with each major city in the world. This will provide an exciting business opportunity for firms located in each region.

Not only will the firms use these communications capabilities, but this technology will enable each of the residents of these new cities to enjoy all the advantages of instantaneous communication to help facilitate their daily operations. As the conceptual plan is implemented, it becomes apparent that many new firms want to position themselves in these new cities due to this ready access to instantaneous information. An extremely competitive environment is created. IIG is a spin-off firm from the early Canadian Trans Canada "data-handling" Network that had been launched in the early '60s to aggressively provide economical data transport across the continent. As a pioneer in the field, the company collected a vast amount of information movement and management experience. In 2005, the spin-off firm was acquired by the progressive RBOC to help implement its two-step program by providing the global networking knowledge to construct the Global VAN . . .

As time passed, the new, planned cities were built near each major city. They were designed so there would be minimum commute time between residences and the workplace. Fiber optic cable was deployed to every home and to every desktop in satellite work centers. New switching systems and transport controllers ensured that both security and survivability were achieved, as well as network integrity and personal privacy. Information exchange enabled the multimedia flow of voice, data, text, image, graphic, and video conversations. There was now an exciting array of new customer-premise equipment, as well as numerous specialized databases to provide financial, legal, medical, and architectural information, and information on protection, education, entertainment, and sports.

It was indeed an exciting change in personal lifestyles. As the cities were being interconnected, it was now time, in 2010, to interconnect them to the major cities throughout the world and, selectively, to the rising, blossoming third world. More and more countries began to extensively embrace various forms of information handling.

Many firms began locating in the new cities in order to provide America with exciting new products, growth, and revenue opportunities. Each industry was being reassessed as to what new products could be made in America, now that these cities had full communication capabilities. It was becoming evident that the next task was to complete the restructuring of all American cities. This was the beginning of a shift back to American product competitiveness. These new multinational firms returned attention to America as they further expanded globally. Yes, it was the beginning of Global Economic Wars, but it was now time for America to participate. It had remained too long in a slow-growth period that had been established in the '90s, while its population increased and manufacturing capabilities dwindled.

During the late '90s and first decade of the new century, some overlay networks had been established in selected areas. Some American firms remained viable, but now there was an opportunity to fully participate in the global arena in an extensive, competitive manner. This was especially true now that the global VANs were being implemented.

So, we leave you with the video conference call, where planners and decision makers discuss the new opportunities provided by merging the new, sophisticated, local communication networks with the global networks, enabling cities to become networked together in a global grid with instantaneous, versatile, economical, multimedia, "information-handling" communication capabilities. (See FIG. 5-2.)

What are the opportunities of this emerging Global Information Society in this Age of Information? What will be the opportunities over the next 1,000 years, in this new millennium, the Information Millennium, the new frontier? The "society-technology-technology-society" cycle is now ready for its next step. The technology is being placed. What type of society will result? Perhaps you will be part of that videophone conference. What opportunities do you

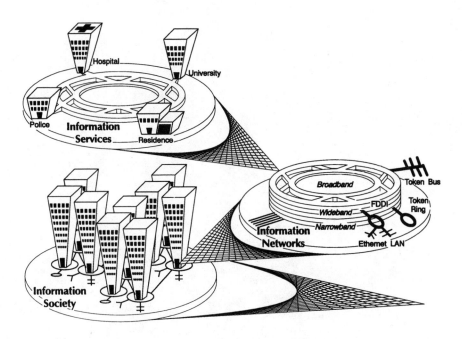

Fig. 5-2. The Information Millennium—the Global Information Society.

think will be discussed . . . as you arise for the early call in Seattle or finish a day in Geneva—or somewhere else around the globe?

"That's very curious,'
she thought. 'But
everything's curious today'" . . .

Carroll

Epilogue

Restructuring America's telecommunications infrastructure

"A long journey I have begun,
in taking the first step,
and proceeding along,
a step at a time,
a step at a time . . ."

Robert Frost

As business leaders throughout America wrestle with the short-term operational goals of today's marketplace, it's essential to keep long-term goals and directions in proper focus. With this in mind, let's consider several steps that the American communications industry could take as it attempts to establish the correct evolutionary/revolutionary entrance into the forthcoming "Information Millennium," the "Age of Information." For example, U S WEST Communications spans the American continent from the Mississippi to the Pacific ocean, and the business has the vision to be one of the finest companies in the world by connecting their customers to the rest of the globe by the year 2000. What steps could telecommunications firms take to achieve the challenging objective of restructuring America's telecommunications infrastructure?

A five-step program: Network Service Plan 2000

Step 1. Public switched voice network services.
Step 2. Private overlay networking services.
Step 3. Public switched data network services.
Step 4. Public switched wideband network services.
Step 5. Public switched broadband network services.

It's important to keep in perspective the tremendous investment in current plant, facilities, operations, and personnel capabilities; it's also crucial to remember the current needs of the customers—the users. As noted in my book, *Global Telecommunications*, there are several user groups, ranging from narrowband to broadband, who require new communication network services. There are also pressing financial concerns, in terms of the depreciation rates of existing systems, as the network moves from analog to digital transport capabilities. The bulk of the existing network is copper based (not fiber). It's extremely expensive to change everything immediately. There will be a transition period for both a physical-plant change and for a customer-information-use, mode-of-operation change.

Step one

In step one, the network services for the bulk of the users can be easily upgraded to the advanced voice services noted in TABLE E-1. Analog-to-digital conversion will enable new SS7/ISDN digital voice services such as the CLASS family of services. Here calling-party identification information is provided to the called party, enabling selective call handling such as specific call transfer to a customer's car phone, call blockage, priority override, calling-name display, high-quality transport, second voice line, voice mail, and voice-grade fax.

Table E-1. Step One

Public Switched Voice Network Services
1985 – 2000

- Telephone
- Dialtone
- Custom calling
- Centrex/centron
- Application service centers
  - Fax (dial up)
  - Voice mail
- Digital upgrade
  - IDN – Integrated digital network
    - Rural clusters
    - Urban overlay
    - Maintenance/administration
    - Digital interoffice
      - 1 error in 10^7 bit/sec

- ISDN – Integrated Services Digital Network
  - Voice/audio services
    - 2nd voice line
    - High quality voice
    - Stereo 7 Khz audio
    - "D" channel signalling
  - SS7 signalling – CCITT signalling system seven
    - CLASS
- OAM&P upgrade
  - Distributed administration
  - Distributed operation
  - Distributed maintenance
  - Automated service orders
  - Rapid provisioning
  - Rapid testing

Step two

It's extremely important that the public provider meet the needs of the private networks. A step towards private-public internetworking is an overlay network that uses public transport but achieves private-to-private interfaces via specialized platforms that tailor solutions for today's LAN-to-LAN interconnect needs. In this manner, private networking can be more quickly facilitated, using systems that need not be fully ubiquitous, functional, standard, or versatile. These more public-network-oriented capabilities and availabilities will take considerably longer to develop and deploy.

It's important that immediate needs be met while future needs are being resolved. With this objective, step two encourages an immediate response to today's private network interconnection needs. Provided, on a selected basis, is a dynamic bandwidth, initially limited to the wideband range, but available where most needed. This offering can then be followed by a more detailed program that enables broadband capabilities on a more selectively shared basis, as noted in the subsequent steps. (See TABLE E-2.)

Step three

Why isn't the deployment of the Public Data Network begun in step two? In one sense, it's as if both steps two and three are performed in parallel. Where the private networking solutions of step two can be provided immediately, the fully operational solutions of a ubiquitous public data network will take several years to develop and implement. Indeed, initial "data-networking" services can

Table E-2. Step Two

Private Overlay Networking Services
1991 – 2000

- LAN to LAN interconnection
  - Frame relay
  - FDDI-I
  - FDDI-II
  - SMDS – wideband
  - SMDS – broadband
- Point to point
  - SONET transport
- Point to multipoint
  - Connection-less
  - Connection-oriented
- T1 – T3 hubbing
- Video conference centers (private)

become rapidly available in selected locations by using current public IDN (Integrated Digital Network) systems with additional ISDN data-handling capabilities. However, it will take time to build up ubiquitous deployment.

Where private networking strength is in selective offerings, public network strength is in standardized, ubiquitous offerings. This takes time and money.

Also, with the more ubiquitous objectives comes complex and extensive address and routing requirements. Therefore, any terminal can reach any terminal, anywhere, anytime. This leaves many technical issues to be resolved. National ISDN data-handling interconnect studies need to be finalized. Traffic blockages must be eliminated. Success encourages more and more usage, as open network architecture interfaces are established to enable multiple providers of multiple services. Therefore, step three is the key step for launching the narrowband data-handling services that will initially satisfy large numbers of data users' needs (as identified in chapter 3). This positions the marketplace to later advance these users to higher bandwidth services, as they move from data and text to image and video applications. (See TABLES E-3A and E-3B.)

Table E-3A. Step Three

Public Switched Data Network Services
1992 – 1995

- Services
  - Data dialtone
  - Dataphone
  - Imagephone I
  - Videophone I
  - Data addressing
  - E-Mail
  - Broadcast
  - Polling
  - Data digital fax group 4
  - Delayed delivery/network retry
  - Security
    - Encryption
    - Audit trails
    - Terminal access
    - Terminal ID
  - D channel services
    - Alarm monitoring
    - Energy management
    - Meter reading
    - Home incarceration
    - Weather sensing
    - CPE to network signalling
    - ESP to CPE signalling
    - Terminal ID

- Services (continued)
  - Integrity
  - Privacy
  - Multiple data base access
  - Closed user group
  - Third party software
  - Record buildup via polling
  - Classes of service
  - Access networks
    - IXC – data networks
    - ESPs
    - ISPs
    - Global data VANs
  - Customer controlled VPN
    - Dynamically
    - Per call
    - Per day
  - Application service center services – access
    - Data storage/retrieval
    - Data manipulation
    - Data presentation
  - Class of service
  - Pricing
  - Billing

Public Switched Data Network Services
1992 – 1995

- Network
  - Narrowband – ISDN
    - Existing copper
    - IDN-ISDN digital overlay
      - Rural
      - Urban
      - Suburban
    - Circuit switching (64K – 128K b/s)
    - Packet Switching (9.6K – 64K – 128K b/s)
    - Packet assembly/disassembly
    - Ubiquitous
    - Information transport services
      - Protocol conversion
      - Code conversion

- Network
  - Narrowband – ISDN (continued)
    - Alternate routing/attempt limits
    - Guaranteed error rate throughput
    - Data rate interfacing
    - Multi rate interfacing
    - Blockage/reroute
    - Message sequencing
  - Public data networks (Ckt, Pkt)
    - Addressing
    - Routing
    - Traffic loading
    - Terminal interfaces
    - Network interfaces
    - Provisioning
    - Test/support

Step four

As the private networks are selectively overlaid and the public data networks become developed and more universally deployed, it's time to turn to establishing wideband/broadband access nodes in more ubiquitous locations. This will greatly expand the offerings of step two, and provide more robust channel-switching capabilities that enhance and expand virtual private networking over the public facilities.

It should also be noted that these platforms will be the wideband/broadband switching nodes of the '90s. They're key to establishing the initial front-end broadband services interfaces with customers that will later be enhanced, expanded, and used by the future, new, fully broadband switches that will be available at the turn of the century.

These initial platforms will be important in establishing the new fiber deployment topologies that achieve fiber to the home, fiber to the desk, and fiber to the conference table. Customers will use more and more bandwidth to fully integrate continuous-bit-rate streams of voice and video offerings with the more bursty variable-bit-rate data. (See TABLE E-4.)

Step five

Publicly switched broadband network services, as noted in TABLE E-5, will parallel step four and step two, as users request more and more economical bandwidth. Videophone needs to be available on a services basis, not on bandwidth basis. As these greater bandwidth offerings are no longer priced on the basis

Table E-4. Step Four

Public Switched Wideband Network
1993 – 1997

- Access nodes
- Fiber deployment (selected)
- Computer to computer
- Virtual private networking
- Closed user groups
- Variable "channel" bandwidth – multi rate
- Wideband data address
- Protocol interface
  - Ethernet
  - Token ring
  - Token bus
  - FDDI
  - Frame relay
  - SMDS
- Videophone II
- Imagephone II

- Customer variable bandwidth control
  - Set up
  - Dynamically
- IXC network interfaces
- P&P internetworking
- Wideband service ESPs access
- Fiber switched networking
  - Class 5
  - Class 6
    - Sub 6
- Wideband OAM&P
  - Phase 1 broadband/wideband
- ATM/STM switching
- Dynamic bandwidth pricing
- Video conference centers

Table E-5. Step Five

Public Switched Broadband Network
1995 – 2015

- Services
  - Video dial
  - Video address
  - Videophone III
  - Workstation (B)
  - Video conference center (public)
  - HDTV
  - Computer networking
    - CPU – CPU
    - CPU – workstation
    - Application service center
  - Video/image
    - Video files
    - Education
    - Sports events
    - Entertainment
    - Search & browse
    - Image storage
    - Image manipulation
    - Image presentation
  - Broadband info switches
    - Closed user groups
    - Global VAN gateways
    - Video storage access

- Services (continued)
  - Video mail
  - Integrated services
    - Narrowband
    - Wideband
    - Broadband
- Network
  - Fiber based – new topology (1990 – 2015)
    - Interoffice (1990 – 95)
    - Selected distribution (1992)
    - General distribution (1993 – 2015)
  - Narrowband-wideband-broadband inclusive
  - SONET/B-ISDN interfaces
    - User network interoffice (UNI)
      - 155 Mb/s
      - 600 Mb/s
    - Network-network interface (NNI)
      - ATM/STM
      - (N x OC-1) switching
  - HDTV distribution
  - Broadband IXC & VAN internetworking

of N number of equivalent voice channels, we'll leave the world of voice-grade offerings, (where video-type offerings require considerable compression and result in flickers of information modulated on low-speed error-prone transports), as we enter a new world of high-resolution, high-quality transport. (See TABLE E-5.)

Challenges

As we labor to achieve the new fiber-based information infrastructure, several interesting challenges lie before us. Further challenges await in the wireless domain for cellular data transport and personal communication network (PCN) data services. These opportunities encompass a wide range of issues—both marketing and network.

Hence, new pricing structures are as essential to broadband market success as the technical choices of ATM or STM technologies are to its electronic or optical transport mechanisms. Pricing must take into account the full range of narrowband, wideband, and broadband services. This issue is especially sensitive to the correct promotion of new video high-bandwidth intensity services, as well as higher wideband/broadband channel information transport between CPUs.

In order to encourage customer understanding and use of the Public Data Network, advertising is especially important in step three. So also will be the achievement of the needed technical addressing and routing standards for ubiquitous interconnection. Similarly, efficient and effective traffic loading and network management will be essential to handle success in the broadband domain, as the full range of narrowband, wideband and broadband services are multiplexed together on integrated, public, broadband, switched facilities.

Issues

Therefore, to "play the game" successfully, we must address the numerous market, network, financial, and regulatory issues, in order to:

- Achieve ubiquitous deployment in key cities.
  a. Using existing copper plant.
  b. Adding digital switching.
  c. Deploying fiber selectively or ubiquitously.
- Deliver massive advertising.
  a. Educate customers on data-handling applications.
  b. Show customers the full range of data-handling solutions.
  c. Encourage customers to use more and more data-handling communications.
- Price for ubiquitous use.

- Formulate partnerships between switch and CPE suppliers and network providers.
- Resolve regulatory/nonregulatory services.
- Enable multiple ESP/ISP service offerings.
- Determine ONA/AIN/INA standards.
- Establish geographic time frame for rural/urban offerings.
  a. Narrowband.
  b. Wideband.
  c. Broadband.

Service nodes

It should be noted that parallel with each of those steps are the extended service platforms, which offer customers many enhanced, shared, specific offerings to ensure that all the information-handling needs are achieved. See appendix A for the full range of potential services and the type of service nodes that can achieve them.

As noted, there are many issues, from deployment strategies for key cities where services are ubiquitously needed, to areas where services are only selectively requested. Here, advertising, price, new plant, old plant, copper or fiber, time frame for phased offerings, standards, regulatory agreements, competitive interfaces, and partnerships all play their role in establishing these new supporting infrastructures for "America's information highways" to its businesses and homes.

Goals and objectives

As America considers the opportunity surrounding the technical possibilities of narrowband/wideband/broadband services, it will be extremely important to establish feasible objectives. It's important to say we'll have fiber to every home by 2015 or 2050 or . . . , but it's equally important to recognize that there are several games to play and several ways to play the game. For indeed, there's the narrowband game, the wideband game, as well as the broadband game. We can begin playing all three games today with different time-frame objectives for different accomplishments, as noted in TABLE E-6. They're only provided as figurative examples of making goals and objectives for future information services more specific, more personal, perhaps more realistic! (Generally, each sector's industries will need services from all the networks; these examples are simply selected for reference.)

Five actions

Five actions will be needed to help achieve the five steps. Providers, working

Table E-6. Narrowband, Wideband, Broadband

Narrowband Services

Narrowband - Copper
64K b/s

Voice messaging, CLASS, Selected Messaging,Ckt Switching, Packet Switching Protocol Conversion, Error Rate Control, Alternate Routing, Priority Messaging, E-Mail, Delayed Delivery, Voice/Text, Data Base Access, Encryption, Audit Trials, Broadcast, Polling . . .

Data Networks Stock Exchange Networks Police/ FBI Networks State Agency Networks Auto Parts Networks Auto License Networks Inventory Control Networks

Financial Legal Small Business Residential

Example Goals;
50% major cities by 1993
20% rural by 1995
70% urban by 1997

Wideband Services

Wideband - Copper - Fiber
(64K) - 1.5M b/s

Bandwidth Management, Wide Area Network, Dynamic Bandwidth, Private to Public Internetworking, ISDN-Non ISDN Internetworking, POP Access, Channel Switching, . . . Survivable Private/Public information Intermetworking

X-Rays, Patient Records CAD/CAM Networks Graphic Display Networks Computer to Computer Networks 1.5M b/s Picturephone Networks Video Conference Networks Wide Area Networks

Medical State Manufacturing Securities

Example Goals:*
20% major cities by 1993
10% rural by 1995
50% urban by 1997

Broadband Services

Broadband -Fiber
N (50M b/s)

Multiples of 50M b/s Switched Transport Video Conferencing, Picturehone, HDTV, Computer to Computer, High Speed Data Transfer, SMDS, FDDI Frame Relay, Call Relay (ATM), & Broadband Information Transfer, Storage, Access, & Presentation

High Resolution Picturephone Entertainment Media Events Education Video Medical Imaging High Definition CAD/ CAM High Definition TV Computer to Computer Data Base Manipulation Visual Presentations

Educatiojn Enterainment Large Business Residential

Example Goals:*
20% major cities by 1995
10% rural by 1997
50% urban by 1999

*These example goals are presented as an example of the need for having a plan of action that can be realistically acheived & agreed to by all parties. Each LEC must establish this type of program in order for suppliers & users to adequately prepare for these type of offerings.

with their customers and their equipment suppliers, need to plan, lead, organize and control by:

1. Establishing an overall program for each of the five steps.
   - Vision.
   - Goals and objectives.
   - Plan of action.
2. Obtain "buy in" by the regulators, financiers, suppliers, stockholders, and the providers themselves, who will be furnishing the offerings.
3. Organize appropriately to achieve success—in both planning and implementation.
4. Phase and launch the "planning activities," working with user groups to ensure that the market requirements truly reflect the users' needs. Be careful to teach and explain to the users the full magnitude of the technical possibilities.
5. Phase and launch the "implementation activities," with checkpoints to ensure that key issues are resolved. Pricing must be for growth. Services must be available in a timely fashion. The network must have the desired degree of quality operation in order to obtain the highest degree of customer satisfaction.

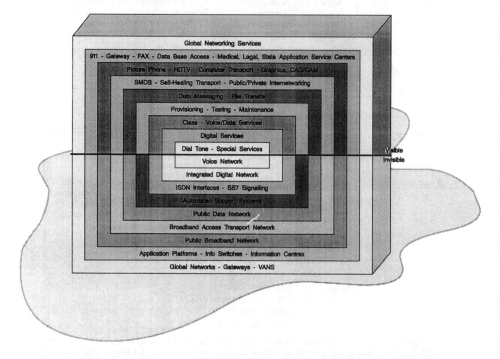

Fig. E-1. The global information network services infrastructure.

Conclusion

As we take these five steps towards achieving the desired programs for restructuring America's telecommunications, we need to occasionally pause and reassess where we're going. Are our five actions being implemented successfully? What other actions are needed? Are our customers responding to our marketing approach? What new advances are being achieved with new technology that can shorten our path? This is a new frontier. It's a time of new challenges (see FIG. E-1). It will indeed be a time of growth and expansion, which is always both challenging and exciting, as we hasten to enter the somewhat dimly lighted portal into the dawn of the new Global Information Society of the Information Millennium.

> The shapeless image,
> changes and changes again,
> to assume form and structure.

Postscript

*"I have sent my heroine straight
down a rabbit-hole . . .
without the least idea what was to
happen afterwards."*

**Charles Ludwig Dodgson
"Lewis Carroll"
July 4, 1862**

This completes the series on telecommunications planning. We've moved from Integrated Services Digital Networks (ISDN) to Integrated Networks' Integrated Services (INIS), to Layered Networks' Layered Services (LNLS). We've considered new processes, technologies, networks, features, services, applications, and societies via *Telecomunications Management Planning, ISDN in the Information Marketplace, Global Telecommunications,* and *Telecommunications Information Networking.*

I hope these books have been as interesting and enjoyable for you as they have been for me. I hope I've "tickled your mind" with future technical possibilities that can help meet future market opportunities; I also hope these thoughts will help achieve a new telecommunications infrastructure that will assist society in its entrance into the Age of Information. We've already progressed through the Age of Darkness, the Age of Reason, the Age of Mathematics, and the Age of Inventions, and, hopefully, the ideas contained in these books will help support civilization as it continues to advance into the Age of Information, and on to the Age of Understanding and Wisdom; there, I hope we'll use these new technologies and services, together with the morals and values of our religions, to obtain a higher standard of living, a better quality of life, and a better relationship with our planet, our neighbors, our family, and our God.

Adieu,

Robert K. Heldman

Part IV

The
information decade
A crossroad in time
The '90s

The time is now.
Now is the time . . .

Appendix A

Private & public internetworking

"Too late to get another boat . . .
We're not going to win the race . . .
sailing the same sail . . .
We're not going to catch up
sailing what they're sailing. . . ."

Dennis Conner
American Cup race

As new narrowband, wideband, and broadband networks are established over the '90s, communications companies need to introduce a new network topology—one in which there are new switching nodes closer to the user. This new topology will take advantage of the fiber and facilitate private and public internetworking. Earlier works have noted the need for a Layered Networks' Layered Services hierarchy of networks and service nodes to accomplish these objectives. With a new topology, a new Class 6 node takes advantage of the robust fiber to promote private/public internetworking by providing access points on survivable rings. These nodes serve as ring switches that offer the full range of interconnect services. These switches are Class level because they provide address translations and routing to different switching and service platforms, be they an interexchange carrier Point of Presence (POP), Value Added Network (VAN), Alternative Network Service Provider (ANSP) or the Local Exchange Carrier (LEC) service node's Information Switch (Info Switch), which could be established as a regulated or nonregulated entity.

Information switching systems

As FIG. A-1 indicates, private network nodes can bypass the traditional carriers or access network services via these new Class 6 nodes. Information switches can provide shared or specialized services above the network, with or without customer-premise systems. Extensive internal CPE networking may be established via new Class 7 entities that provide access to traditional Class 5s, POPs, or service nodes and info switches.

Class 6 switch—What and why?

In considering a new Class-level switching node located closer to the customer, we have to ask the questions why, when, where, and, of course, what and how. When asked what's accomplished with the Class 6 switch, we might consider the following:

- Obtain survivable transport.
  a. Home on multiple base units.
  b. Survivable rings.
- Use the fiber to its maximum.
  a. Share facilities as close as possible to the customer.
- Reduce mileage cost to the customer.
- Enable direct access to other carriers' Points of Presence (POPs), without going through the entire public network.
- Achieve Virtual Private Networking (VPN) under customer control.
- Enable dynamic bandwidth allocation.
  a. Multirate—n <x> 64K b/s
- Provide narrowband ISDN/non-ISDN networking interface.
- Provide wideband/broadband interfaces.
  a. Frame relay interface.
  b. SMDS interface.
  c. FDDI interface.
  d. ATM interface.
  e. STM interface.
  f. Channel-switching interface.
  g. Network management interface.
- Provide P&P (private & public) internetworking access point.
- Enable shared transport between nodes for private networking over publicly shared facilities.
- Provide quick access to info switch service nodes for advanced services.
- Control Sub 6 distribution plan fan-out for Fiber To The Home or Pedestal (FTTH-FTTP).
  a. New fiber distribution plan.
- Provide interface node for broadcast capabilities of High-Definition TV (HDTV) and/or Private Communications Services (PCS) network via Class 6 or Sub 6.

- Provide front-end translations and interfaces for Layered Networks' Layered Services addressing.
- Provide direct access to multiple Class 5s or, via multiple Class 6s, to multiple Class 5s (Note: This is a key reason for Class-level access.)
- Provide front-end functions to interface over INA/ONA to a new Class 5 broadband superswitch, other Class 6s (to prevent "ring around the rosie"), and the new customer-premise Class 7 switch; also provide INA/ONA network interface functions to interface with info switch-type service nodes.

Class 7 switch—What and why?

Over the years, many have questioned the difference between a PBX (private branch exchange) and a PAX (private automated exchange). The answer was quickly forthcoming and easy to understand: With the PBX, a customer can originate and receive calls from the network, while a PAX is strictly an internal system on customer premises. While PAXes usually don't have an internal operator, PBXes usually have an operator to intercept calls from the outside world and route them to an internal address. More sophisticated PBX fifth-generation systems have voice answering capabilities that automatically ask

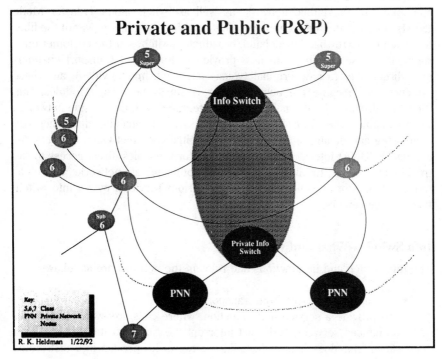

Fig. A-1. P&P.

the outside party to dial the internal address, once the network is finished completing the call to the PBX.

Centrex/Centron-type services were initially provided on a network basis to closed user groups, where numerous customer campuses could obtain services from shared platforms with other user groups. Here, internal numbers were made available to the outside. Their terminals could be reached using a hunting mechanism that searched through available access lines. However, rather than use a single common number to the system, the individual internal number could be provided without requiring the intercept operator or announcement for inward dialing.

With the advent of data came LANs, cluster controllers, distributed network gateways, etc., to internet data networks within the customer premises or between customer premises on a wide area basis (WANs) or around a city as metropolitan area networks (MANs).

With the discovery of economical T1 (1.5M b/s) and T3 (45M b/s) transport capabilities came the private-to-private (inter) networking nodes (PNNs), located on selected customer premises, using leased facilities between nodes. Here, protocol conversions, bridges, routers, and gateways blossomed to achieve LAN-to-LAN internetworking, as Frame Relay and FDDI protocols emerged and blossomed.

However, the need to internetwork more and more terminals to more and more locations focused attention on the Asynchronous Transfer Mode (ATM) technology of the public network, as well as Synchronous Transfer Mode (STM) circuit-switching technology. This combined with the power of the fiber to introduce narrowband/wideband/broadband switches in the customer environment. These switches can now provide n number of transport channels over direct fiber to any terminal (voice, data, text, image, video), and these switches will become the new PBXs or Business Servicing Modules. But, they'll be distributed internally and fully integrated with the outside networks. In this manner, they'll become the new Class 7s, by enabling direct network addressing and dialing to any customer terminal via direct Open Network Architecture (ONA) interfaces. Hence, the platform enables direct access to the traditional network, to internal networks, to other private networks, as well as to other Points of Presence (POPs), or to various service-provider info-switch-type service centers.

Info Switch—What and why?

The Info Switch and its functions and roles in the network are as follows:

- To separate services from transport.
- To provide a service node function, where calls are switched to it for enhanced services. It doesn't interrupt the call control at the transport layer, nor does it provide in-line software; it operates in a distributed processing manner, adding further value to the call.

- It can be accessed either directly from the equipment on the customer premise or via the new access nodes, located closer to the customer (ring switch/Class 6 switch) or directly from private network nodes or from the traditional Class 5s. If located in the regulated network, it can perform CU-Centrex/Centron-type functions, but for a full range of information (voice, data, video, image, functions), not just voice.
- If located in the unregulated/separate subsidiary arena, depending on further Information Services rulings—it can contain selected third-party software, sharing usage with numerous partners.
- The key to the Information Switch over previous systems is that it's both a central service serving system from both a transport and a database file management perspective (half C&C); it also uses CPE terminals, systems, and subsystems that can appropriately collect and distribute information on customer premises, using standard ONA interfaces between the CPE and the switch—thereby enabling direct access to the Public Transport Network from CPE to reach the info switch.
- Features and services—The info switch can be several different systems, since it will operate in all three transport media—narrowband, wideband, and broadband. It can provide services to closed user groups; it can serve as a "private-to-public" networking node or "private-to-private" networking node for special networks. Or, on the shared transport side, it can handle many communities of interest and enable cross-community-of-interest interconnection on a secured basis. It can provide access to stand-alone application service centers or contain an adjunct service center node for specific services.

Some have visioned using an info switch to serve the entire medical community in a major city; others would use it as an Information Centrex, serving several closed user groups of different communities of interest. It can interface with both sophisticated CPE systems, such as LANs or Class 7 broadband switches, as well as singular terminals or mainframes.

On the computer side, its service node can perform sophisticated data storage, access, manipulation, and presentation functions, using high-level, friendly, user-to-machine interfaces so the customer can point, touch, or talk to very sophisticated terminals. Or, on the transport side, it can perform sophisticated security, integrity, and transmit functions, such as: deny access, encrypt, audit, password, ISDN/non-ISDN interface, error correction/detection, dynamically variable bandwidth via customer control, VPN, record building, polling, broadcast, addressing, routing, search, file, delayed delivery, and store and forward messaging (voice, data, video, image) for both the residential and business (large and small) customer. In time, access mechanics will be available through INA (Information Network Architecture) to change via ONA transport systems databases to enable selected service routing to specialized private Info Switches. In addition, "D" channel signalling paths will enable transparent transport communication between the CPE and the info switch through the traditional network. Finally, direct access can be achieved from the info switch to IXCs and VANs, serving as a gateway to POPs or by re-establishing the call or a sequence of calls back to the Public Network—for example, to call all the hospitals in the area to determine if a particular person was a patient, in order to provide all available X-rays to a local or distant specialist.

In this manner the Info Switch serves as a versatile multimedia handling system. Most suppliers have noted that it could also serve on customer premises as a private switching node for the customer's communication complex (Class 7)—especially if its internal design is modular to the point of also being physically distributed throughout the customer's campus.

These general thoughts denote (See FIG. A-2) the possibilities of such versatile additions to the network, using the Layered Networks' Layered Services model to denote ONA location, as well as INA interface, and function. In this manner, we'll be able to achieve narrowband services over existing facilities, overlaid with conditioned transport wideband services, and then augmented by a growing array of fiber-based broadband services to formulate an expanding and encompassing Broadband Network Services Infrastructure.

> *"Those who explore an unknown world are*
> *travelers without a map; the map is the result*
> *of the exploration. The position of their*
> *destination is not known to them, and the*
> *direct path that leads to it is not yet made."*
>
> **Hideki Yukawa,**
> **Japanese physicist**

Open Network Architecture (ONA)

One might step back and ask, "What is ONA? What was it meant to be? What is it becoming? What should it be?" At divestiture, under Computer Inquiry II (CI-II), the FCC ruled that the BOCs could offer enhanced services and CPE only through separate subsidiaries. Then, in 1987, the FCC relaxed the separate subsidiaries rules for CPE. Later, the FCC adapted the Computer Inquiry III (CI-III) Open Network Architecture rules for enhanced services. In early 1988, RBOCs enhanced their original ONA plan, obtaining interim approval for voice messaging and protocol conversion. The interim plans provided for CEI (Comparably Efficient Interconnection). In June 1990, the U.S. Court of Appeals eliminated the CI-III rules, but still permitted the RBOCs to continue to offer integrated services subject to approved CEI plans, but required that all other enhanced services be offered through a CI-II subsidiary, much as it had through divestiture. Here, the FCC readapted all of ONA except those parts dealing with the elimination of structural separation. As a result, CPE may be offered on an integrated basis, subject to nonstructural safeguards. An RBOC may offer certain limited enhanced services on an integrated basis, subject to approved CEI plans. Enhanced service is data or information that's carried

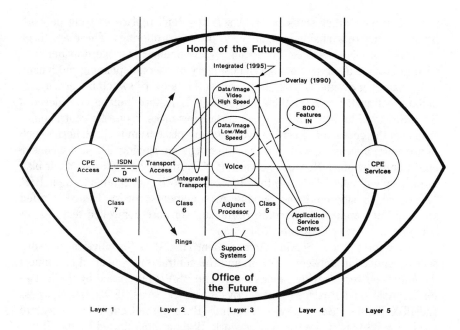

Fig. A-2. An eye to the future—layered networks' layered services.

over a network and changed, sorted or modified. For example: voice messaging is an enhanced service because it stores messages on incoming calls. Other examples of enhanced services include electronic mail and facsimile storage. Some RBOCs distinguish between external Enhanced Services Providers (ESP) and its own Enhanced Service Operations (ESO). CPE, of course, is simply communications equipment at the customer premise, such as private branch exchanges (PBXs). The only exceptions are for multiplexing, voltage protection equipment, inside-wire and central-office-controlled coin phones. On October 7, 1991, the appeals court lifted the ban, which had been opposed upon the BOCs entry into the information service market, thus pending further judicial action; RBOCs are now free to offer full services in the information services market. For purposes of FCC compliance, there are no meaningful differences between enhanced services and information services. Besides, the FCC mandates on ONA as to how the RBOCs can offer their basic network and integrated enhanced services; there are further rules and regulations mandated by the Department of Justice, addressing the Modification of Final Judgment (MFJ), and, for some RBOCs—Civil Enforcement Consent Order (CECO) and the Enforcement Order (EO) requirements, which ensure that they have adequately processed and approved the introduction of new services and service trials. (See universal information services in the glossary.)

ONA services are network features, functions or access arrangements that are needed in order to provide an enhanced service. ONA services are unbundled, meaning that each service can be bought separately without having to purchase other services. ONA services don't replace existing network products; they're available in addition to packaged offerings. There are three categories of ONA services. For example, an enhanced service provider may wish to provide a service to a special category of users, depending on dynamically changing events. Hence, they request the network switch in certain conditions to transfer calls to their platform. This feature can be considered a Basic Serving Element. They may, in certain instances, request a central office to inform the customer of certain events taking place from their platform, such as providing a message-waiting indication or custom ringing. These are considered Complementary Network Services (CNS). In moving the call to their platform, they will most likely use the physical network connections to their ESP. These are considered Basic Serving Arrangements (BSA). Hence, tariffed physical interconnections become BSAs, and software controlled features become Basic Serving Elements (BSE).

Therefore, "Basic Serving Arrangement" (BSA) is the physical network access connection between the ESP and the central office, needed to connect the enhanced service with the customer. The BSA is purchased by the ESP—for example: Flat Business Lines, Private Line Services, ISDN Basic Access and PBX Trunks. "Basic Service Element" (BSE) is the Central Office feature used with BSAs to make the service work. BSEs are purchased by the ESP—for example: Dial Call Waiting, Hunting, Call Transfer, UCD and Market Expansion Line. "Complementary Network Service" (CNS) is the Central Office features that are used on the ESP end-user's line and are needed to work with the enhanced service. CNSs can be purchased by the end user or the ESP for the end user—for example: Call Forwarding Services, Custom Ringing, Expanded Answer, Message Waiting Indication and Speed Calling.

Implementation rulings say that if an RBOC uses basic network services (access arrangement or features) as part of its enhanced service, basic services must be made available to all customers on equal terms and conditions. Equal terms and conditions must apply to price, installation intervals, maintenance intervals, availability and functionality. When an RBOC uses an ONA service as part of its own enhanced service offerings, the ESO must pay the same price for the use of that ONA service, as any other RBOC customer.

If an ESP requests a basic network service for its enhanced service and the request meets specific rules established by the FCC, the RBOC must provide the service. The FCC requests that the ONA services be unbundled; that is, they must be offered as an individual network component with specific rates. All ONA services must: not have usage or user restrictions, be available for use by all customers at least by the same date they're used by any RBOC enhanced service, be available to all ESPs for testing prior to the date the ONA service is available with or as part of an RBOC service offering, and be developed to meet the identified needs of all enhanced service providers. An RBOC's ESO can't use packaged basic network features as part of their enhanced service offerings; Centrex/Centron services have not been classified as an ONA service and therefore can't be used with the RBOC's enhanced services. However, certain features such as call forwarding and message waiting indication, which also can be provided by the network, can be used with an RBOC's voice messaging under certain circumstances. These limitations of using Centrex/Centron features with voice messaging apply only with the RBOC's enhanced services offerings and don't apply with ESP offering enhanced services, but for Centrex/Centron customers, this can be a complex matter.

Observations and conclusions

In looking at the above rules, regulations, and reservations, it's quite apparent that as the system is opened for rapid service delivery or advanced intelligent networking, Pandora's box may indeed be opened. The more anyone is able to access any point in the internal network switches, the more dangerous and complex it becomes from both an operational, administrative, and regulative point of view. As the industry enters a world of intense competition for providing telecommunications to anyone, anywhere, anytime, the network tries to open without destruction to accommodate multiple providers of multiple services. This has been the desire of Judge Greene for the past many years since divestiture, and it has become the battle cry of the new competitors as they demand equal access to local customers. Colocation of the systems, portability of numbers, and direct interconnection to every possible aspect of the network has been their objective. Unfortunately, earlier BOC resistance to the Carterphone-type CPE hookups to the network has weakened the RBOC's claim that it can't be done without bringing the network down. Similar resistance by the large computer firms, such as IBM, to limit direct interfaces to their mainframes has led to open system demands in the computer world to provide direct interfacing to other equipment manufacturers (OEM) systems and universal protocols that enable the sharing of information.

However, within the communications industry, this cry for openness has led some within the RBOCs to promote direct access to each step of the transport process, hoping to select each function as a marketable entity and, at the same time, to meet the FCC demand for openness. Unfortunately, seriously looking at the ramifications of opening one's home so anyone can walk into it, use it, live in it, make a mess in it, is something else indeed. One may be able to accommodate a few visitors, but thousands and thousands may be overwhelming. What you could do with a few may be totally impossible with many. Even Washington found that, when they attempted to open up CCITT study groups to everyone, crowds of 400 and 500 interested parties could not fit around the planning table. There was no place to put them, no way to meet all their conflicting demands; soon, they gave up and returned to more controllable audience sizes, using mechanisms, such as T1 committees.

As we look at the goal of openness, we can't disagree, but as we look at how this openness could be achieved, we might ask, does using 20 pages or so to describe the call transfer operation really achieve an obtainable service?

As more and more features are needed and used by the enhanced service providers, how much should we open up every element of the network, especially as we move into not only voice handling, but also data and video handling. Can this openness be managed successfully from a network perspective, so the network will not go down? It took ISDN until 1992, 10 years, to begin to achieve interconnectibility of different suppliers for the most basic of interfaces. Many believe that the billing, maintenance, and administrative aspects of ONA are much bigger than the hardware and software complexities of this increasingly complex task. With the advent of Advanced Intelligent Networking, where the network operations are interrupted for parallel tasks to be performed by autonomous equipment (via STP/SCP centers above the network), we're in danger now of moving work to hundreds of ESP systems, waiting for correct responses before enabling continued network operations. With the advent of some forms of rapid service delivery, we may be stepping further into the misty bog by enabling alternative software to provide in-line decisions—if this software is provided by hundreds of ESP providers. Not only have we increased the security and survivability complexity of the basic transport systems, but we've also created an administrative nightmare.

So, what's the answer? What could be done to more easily and realistically achieve the degree of openness required to achieve a level playing field for all and still enable the game to be played, or should we continue to attempt to establish unrealistic boundaries, plays, rules, and regulations that inhibit the game?

To play the openness game means to play a game in which the players can win rather than simply run around the field and accomplish nothing. To win requires customers, the users, to be accessible with the minimum number of unnecessary bottlenecks. It also means that we must ensure that each player, including the RBOCs, must be provided a fair-share opportunity to play the game. Care must be taken to ensure that the game can indeed be played; otherwise, the spectator (the customer) is provided nothing.

To do this, networks and services can be layered on top of each other, using the public network's nodes as internetworking points to facilitate private and public internetworking. That doesn't mean every internal element of the transport/switching mechanisms are accessible, but it does indicate that there's indeed the ability to perform distributed processing and handling of their information, using controlled mechanisms at the system level to switch the call from one system to another to obtain parallel transport or higher-level services. To this objective, INA (Information Networking Architecture) by Bellcore will be using the Layered Networks' Layered Services model defined in Global Telecommunications (See Reference) to denote ONA interfaces, which enable the passing of call information from network to network or service node to service node. ISDN d-channel signalling and packet handling can provide specialized intercommunication information to and from the service node and customer by providing the access node, located considerably closer to the customer, and by providing the Class-level features to route the call to a central office or an alternative provider's Point of Presence (POP). The bottleneck is substantially open without disturbing the call processing, by enabling service nodes to be accessed directly from the customer (via Class 7 type CPE equipment), or through network access nodes (Class 6s) located near the customer, or via the traditional central offices (Class 5s) (See FIG. A-1 again.) In so doing, blockage to alternative services is eliminated, with information moving to the desired service platform as rapidly as possible. In time, as protection mechanisms are achieved, access to line-side databases will enable the network switches to perform more of the specialized handling that's desired by alternative service providers, such as: to enable calls to be more quickly routed to their alternative platform for specialized customer handling, or to generate specific requests or responses to and from the customer. How dynamically these databases are updated is yet to be resolved by newer systems that enable databases to become more virtual and transparent to the momentary needs of the call.

Addressing

In summary, these mechanisms are consistent with future distributed processing technology and more realistic than current approaches in their achievement and provision, with less risk to the baseline public network. In this manner, alternate providers can achieve access to their alternate networks and services, as we strive for a reasonable degree of "openness." With regard to addressing and inward dialing, it's obvious that the interconnected networks will require their internal terminals (phones, printers, workstations, etc.) to be addressable from anyone on the network. As these terminals are moved to different nodes on the networks, there's a need for more numbers, as well as a need for sophisticated support systems to quickly update software databases to reflect these moves. Thus, the current ten-digit area-office-terminal codes will need to be expanded in order to accommodate and facilitate the forthcoming multiple networks and multimedia terminals. Here, ISDN has proposed a 15-digit numbering system, which will also accommodate non-ISDN/ISDN internetworking—thereby helping to resolve the future issue of number availability and portability. In the interim, as we overlay networks in order to move a customer to a different serving office or system, the customer will usually require a new office number.

Colocation

Colocation, on the other hand, is a different story. We've seen severe problems in the past where multiple systems are clustered together from different vendors. One system's AC/DC converters failed, causing improper voltages to ride on existing power systems, affecting other systems. Shared emergency generators can be unduly overloaded. Grounding buses have contained unwarranted "noise," as high as 6 volts, thereby totally disrupting the operation of the entire complex. Or, radio frequency interference (RFI) generated from one system can totally disrupt unshielded systems within the complex. The cost for these systems to provide additional protection should not be underestimated in comparison to simply routing the traffic to alternate buildings that house additional systems, thereby allowing each to obtain their desired power, grounding and space for future growth. To accomplish this, traffic could be routed at no cost to these alternative sites within a reasonable area. In our zeal for openness, we could become so open that we're put in situations that are not only dangerous, but also detrimental to the health and welfare of the entire network. Those who push these endeavors should be forced to pay for the additional protection expenditures and the liabilities in the event of network collapse. Today's technology can provide certain degrees of openness. It can provide access and versatility, but this doesn't require stacking equipment on top of each other.

Pricing

As we open up each element of the call flow with separate pricing, we could be creating a nightmare of layering costs on the call flow, which may drive public customers to less-expensive, unprotected private networks (bundled) for the majority of their transport; customers may only use the public unbundled segments (which become cumulatively expensive) as a last resort. This leaves some in the best of both worlds, and the public provider in the worst. Also, one might ask, "Who pays for the expense of providing such accessibility everywhere?" It's the rural areas and less-desirable urban sectors—the nonbusiness districts. In actual usage, competitors only want access in the more lucrative areas to deliver only the more lucrative services in those places.

Segment pricing will also affect RBOCs' parallel private overlays, as they'll be put at a disadvantage as they attempt to use the common public network. Their total service pricing will not be competitive. This then not only drives the large competitor to separate networks from the public, but also diminishes the RBOC's public network from being used appropriately for private offerings. As the public network loses the large players, it thereby becomes a network that's required to carry less traffic than it should economically, leaving a smaller base to pay for the overall internetworking costs. Hence, one can question the common good, especially as the RBOCs become ineffective in using the public network themselves, as they attempt to capture (recapture lost) usage by private networking on top of the public network. This was one of the strengths of the old CCSA-type networks, where transport was shared and switching was provided by piggyback switches.

This split will widen until the public network, carrying the heavy burden of full ONA, eventually collapses by its own weight, with no real revenues to sustain or support the expenditures for openness. As one economist once said, "there's no such thing as a free lunch."

AIN/INA/LNLS-ONA

As we look at the impact of AIN (Advanced Intelligent Network) on ONA (Open Network Architecture), we see that opening the switch in the central office to interrupt, to provide AIN features, also opens it up to similar AIN interfaces for ESP (Enhanced Service Providers), as they wish to offer extensive database look-up services. Next, in time, no longer will many ESP providers be willing to simply ask the RBOCs for their BSE elements to perform services for them, such as switching calls to their platform or informing the CPE side of various indications. They may wish to step into the world of performing inline transport call-handling functions themselves, with rapid service delivery, using an adjunct processor to provide actual inline software. The problem is further compounded as multiple providers begin inserting their software in the midst of network transport software. This indeed becomes a dangerous situation. Therefore, INA (Information Network Architecture) has focused on simply moving the call to the desired platform for further services by tying with the call a contract on what needs to be performed at that location. This then pursues the concept of switching the call to the service node for additional work and then returning the call back to the transport for delivery. Using these forms of interfaces, ONA begins to become more palatable and less dangerous. Finally, using LNLS (Layered Network's Layered Services) INA concepts, new switching nodes are placed closer to the customer, enabling closer direct access from the customer to the ESP provider. In this manner, bottleneck and blockage to the ESP are further reduced, and ESP systems are provided access that's autonomous from central office interrupts. There can still be a reasonable set of BSAs and BSEs without requiring extensive impact on transport switching architecture. As time progresses, when protection mechanisms for dynamically updating databases are available, there may be an opportunity, on a controlled basis, to enabling ESPs to dynamically redirect calls to their systems on the request of the customer.

All in all, allowing numerous AIN interrupts (from many ESPs) to the system can become, traffic-wise, dangerous; adding software inline via RSDs may be more dangerous. On the other hand, switching the call to the service node provided by the ESP is less dangerous. Similarly, breaking up the call flow into too many little segments may provide an inline expense, which can accumulate to the point that it's more favorable to route the calls in a bundled manner via private-to-private networking. Finally, paying the administrative costs and providing the unbundling services ubiquitously could become a maintenance and administration cost that "grew and grew until gruesome."

In conclusion

These "thoughts" require a more extensive review of all aspects of the issue; they have been provided to encourage the "if then—then what" scenario-type thinking. If we do this, then what type of world must we live in from this point on? The objective of unbundling to provide a level of openness is noble and desirable, but to realistically achieve it without constraint and control is to be undeniably dangerous. If the network switching system designer doesn't really know what features the switch must perform for anyone, anyplace, anytime, it becomes, in actual implementation, realistically impossible to simply open up the switch in an unbundled manner so anyone can redesign it to do something else, in addition to, etc., without perhaps destroying its internal integrity and security. Also, seeing how simple computer viruses, generated by unscrupulous software hackers, are capable of bringing down complex computer systems, so the network is vulnerable to this stated objective by various terrorist groups.

Over the years, the ONA process has attempted to meet FCC direction, but some wondered what the RBOCs got out of it, as it was not linked to the information services restriction, but to the FCC side of the issue of achieving the competitive arena via subs, unregulated/regulated endeavors, etc., etc., without thinking out the full ramifications of what the net result would be as RBOC people were moved from regulated departments to unregulated departments, to separate subsidiaries, back to unregulated departments with separate accounting, etc., etc. Now, the next attempt is to do the same with the internals of the network switching systems. Unfortunately, machines aren't as pliable as people, and once designed and set in one manner, machines don't readily accommodate this type of change easily, rapidly and inexpensively. They sometimes break down, but, then again, so do dedicated, good people, when change becomes too severe. The key to ONA's success will be in controlling its currently uncontrolled, changing direction.

Remaining questions

1. What are the basic serving elements that are required to be made available by the various RBOCs? Observation: ESPs are requesting preselected or dynamically changing capabilities for their customers to be routed to their service platforms. Initial BSEs have centered on the voice network services. They've not even yet begun to address the data networks BSEs. They further note that they wish to not only add transfer functions to route callers to their platforms, but they also wish to dynamically add their own call-controlled software to each aspect of call flow. This can become quite unmanageable, let alone dangerous. As one telecommunications planner has noted, "An ESP may fail in any of several ways: completely out of service; slow service; sends unintelligible messages to the network switch; sends syntactically correct messages to the switch, but those messages can't be executed; sends valid messages to the switch, but those messages don't correspond to the client's expectations (e.g., wrong number translation). Any of these

conditions can be persistent or intermittent. Who is responsible for detecting such conditions? How can the network inform the ESP and/ or its affected clients of such conditions? What action should the network take to protect itself and its users?"

2. What are the various concerns and requests for dealing with RBOC interfaces to AIN and SS7, in terms of new network architectures to achieve a modular and transparent transport network? Observation: Some don't believe that the RBOCs are creating a platform from which the Information Age can be launched. They believe AIN is being established for a local environment and that history teaches that the RBOC is unlikely to voluntarily evolve to the desired competitive-arena modular architecture. They note that Section 218 of the Communications Act requires that the FCC inform itself regarding the evolution of the public switched network.

The bottom line is that some don't see a plan for removing or reducing exchange carriers' bottleneck power and facilitating use of the network to develop creative services via service dependent/service independent interfaces to users, interfaces that are accessible by ESPs and ISPs using AIN and other intelligent architecture. They further note the confusion in current RBOC plans in missing migration milestones in the road to the "end state." They challenge the design in AIN if it can't protect the network from harm, if it limits their equal access. They also request an accounting of the cost/benefit to upgrade the network to enable them to provide their services. They concur that wide open is not the objective, but the challenge/threat to networks' security doesn't automatically or necessarily dictate the highly restrictive user environment that's contemplated. They note that the AIN's use of the OSI (Seven Layer Internetworking Model of CCITT) is consistent with international standards, but AIN's restriction to limit their access to only Layer 7 and not Layer 5, 6 and 7 would ensure that, as one FCC filing noted, "the ESP has virtually no ability to access and use network functions than any 'ordinary' telephone subscribers."

Others have specifically noted that some RBOCs don't intend to provide a public interface into call processing operations of the switch itself. They note that RBOCs will continue to determine which BSEs are created and which BSEs are not created. They believe that the RBOC's comments confirm their suspicion that RBOCs intend to limit end-user/ESP access to network functionality to the RBOC-defined "service level." They charge that various overseeing committees are incapable of solving long-range "end-state" network concerns. "Deferred to these bodies, they would become hopelessly delayed, and stalled." They believe that the shift from internal supplier "AT&T"-type equipment to the RBOC for bundling and unbundling services and providing application programs will still inhibit third-party software providers from providing service.

In summary

The preceding thoughts concerning ONA can be summarized as follows. The desire to open up a switching system to enable enhanced service providers (ESPs) to competitively provide their services to the customers is indeed a noble objective. However, we must be very careful that what we do doesn't jeopardize a finely balanced structure. AIN (Advanced Information Networking), was mainly designed for voice "enhanced database search" services. It has indeed tickled the interest of many ESPs to also provide similar services on an interrupt basis, but this challenges network integrity and feasibility, as more and more services are provided by more and more ESPs in this manner. Some rapid-service delivery approaches, using adjunct processors to provide inline software, have unfortunately opened the system to another degree of openness, which further causes concern for ensuring the integrity of the system. Also, these endeavors, depending on pricing of segmented offerings, will affect both public and private networking pricing for similar offerings. As RBOCs unbundle the machines, they must place themselves in the "if then, then what" situation to determine if they can indeed live and survive in the new scenarios that take place.

Some service providers have indicated that Bellcore's AIN architecture, and the various RBOC's versions of it, don't provide for ESP involvement and software control of network switches' call processing. Therefore, some service providers have provided the following questions:

- In what direction is AIN leading the industry?
- What is (should be) the future network architecture that enables a competitive marketplace?
- Why can the RBOC have access that we (ESPs) cannot?

There's a view that new services can now be easily, quickly and inexpensively added "inline" to call control either on an interrupt basis or by adding new software to the system in an adjunct processor. So, it's quite natural that some providers would ask for the same interfaces as Basic Serving Arrangements (BSAs), with an internal set of Basic Serving Elements (BSEs) like call transfer to interrupt call processing of the switch to bring the customer to their service platforms, where they can provide enhanced services (ESP) and information services (ISP).

Pressure exists for immediate relief using AIN interrupts and unbundled pricing for numerous BSA and BSE offerings, but there's considerable danger that, as large expenditures occur for ubiquitously unbundling every aspect of the system, the net result is a very expensive underlaying transport mechanism that's extremely difficult to maintain and administer both economically and safely.

Pricing, colocation, number addressing, interface and portability issues are indeed the challenges of a new network architecture, as we shift from public or private networks to public and private internetworking. We must deal with the entire picture and not just a piece, on both an evolutionary and revolutionary basis, in order to resolve the considerations and observations noted.

Until now, the industry has not provided the world with a vision of where we're going and how we want to get there in terms of a feasible architecture for future networks and services that permit a competitive market to prosper and flourish. It's time to lead the way out of this confusion with these concepts for future narrowband, wideband, and broadband networks and services, in terms of the Layered Networks' Layered Services model. The industry is now at a "crossroad in time."

Therefore, this analysis has attempted to look at the consequences of our actions from the different perspectives of what happens to the RBOCs private offerings and public offerings from a network, service-pricing, network integrity, address availability, number portability, colocation, and overall feasibility and integrity point of view, especially as they not only deliver to the world voice services, but enter the more complex domain of data. (See FIG. A-3.)

In the future, the integration of the Layered Networks' Layered Services model with Bellcore's new INA (Information Networking Architecture) should hopefully provide more viable and feasible ONA access points. Here, access nodes become closer to the user to provide direct access to ESPs. INA offers traveling call-handling service requests to distributed processing platforms. Subsequent ESP direct access to customer transport call control databases, as

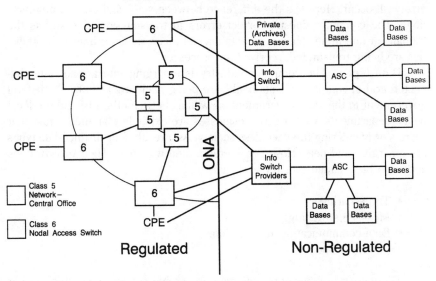

Future Regulated/Non-Regulated Architectural Boundaries Considerations

Fig. A-3. Future ONA boundary considerations.

well as their use of ISDN d-channel communication to CPE equipment directly from ESP service nodes, will enable services to be more transparently provided throughout the Central Office (CO) network.

It might be feasible to open one's house for a few guests to walk around, but when 10,000 are attempting to do so, it's a different story . . .

The global information marketplace today and tomorrow

Soon it will be
that people can live and work
wherever they would,
and be connected with
the whole world.

Global data users

First, for the sake of clarity, let's define what's meant by the various data-handling terminologies.

Data Data is the coded representation of written, typed, or photographed information. It's normally represented in binary form according to some coding scheme of N bits. For example: The American Standard Code for Information Interchange (ASCII) represents the letter R by a 0100101 bit pattern. Hence, in "handling" the letter R (that is, transmitting, receiving, or processing the letter), it's important that this bit pattern doesn't change, or else an error will occur. Here lies the difference between voice and data communications. The computer doesn't correct errors in transmission as easily as the human ear and brain. Therefore, data doesn't mean statistics unless the statistics are being transmitted, received, or processed.

Data types The computer industry is emerging into a giant interconnected grid of computers. Data banks are being made available by the federal government to the state government, as well as between the FBI and the local police departments. Credit card businesses are expanding at such a rate that we're slowly entering the checkless society. As a result, four distinct data types are emerging and need close consideration and watching. Their growth does appear to be dynamic over the early '90s.

- Time-sharing.
- Message switching.
- Data communication/distribution.
- Inquiry/response.

There are a number of remote data processing systems emerging, having different communication requirements:

- Inquiry systems.
- Data collection systems.
- Conversational computer systems.
- Remote batch-processing systems.
- Information distribution systems.
- Interactive graphic systems.
- Remote document-production systems.

Data is being distributed by various methods over private, leased lines or message, direct, switched facilities, depending on the accuracy, speed, availability, and response time requirements.

Data bank A data bank is a large volume of common data stored in an information retrieval system, accessible by local and remote computers and terminals.

Data distribution patterns A number of different patterns are being implemented randomly by various organizations to handle the users on a per-need basis. There's no order or logical neatness to the patterns that are presently being developed. They're costing the customer time and money due to the makeshift interfaces and configurations. The following five types of data distribution patterns are emerging:

- Point to point.
  - a. Computer to computer.
  - b. Headquarters to factory.
  - c. Store to warehouse.
- One to many points.
  - a. Headquarters to sales offices.
  - b. Weather center to broadcast stations.
  - c. Reservation system.
- Many to many points (direct).
  - a. Car rental, city to city.
  - b. Intra-company communications system.
- Many to one point.
  - a. Data collection.
  - b. Stores to warehouse.
  - c. Order entry.
  - d. Credit checking.
- Many to many points (switched).
  - a. Message switching.
  - b. Inter-company transmission.
  - c. Brokerage transactions.
  - d. TWX.
  - e. Telex.
  - f. Fax.

Now that the types of data and their distribution patterns have been indicated, the next logical point is to determine who are the data users.

Data users The major users are probably on the circulation lists of the major computer software, personal computer, and data communications trade magazines. The data users are as follows:

- The governmental agencies of the United States. These are, by far, the majority. These customers use time-sharing for scientific research, data collection and retrieval, inquiry and response, and message switching. The government has built several computer networks over the past twenty years.
- State and local government—police, government administration and maintenance departments, etc. As large data banks are being built and tied between states and the federal government, the state users are increasing their use of inquiry-and-response-type systems. Politicians are becoming interested in dial-up public surveys, etc. Besides serving the large financial processing user (the state and local budget departments), the computers are being given operational control tasks such as traffic control during rush hours for some of the major cities' streets.
- Centralized companies—such as Sears, International Harvester, large department and food-chain stores. Here's where the major increase in time-sharing is expected. As better identification techniques occur and more remote store branches are built, remote entry will increase for a multitude of purposes such as inventory control, verification of customer credit, and warehouse stock distribution.
- Time-sharing service centers. These facilities are being offered to business people who can't afford large in-house computers and staff. Some educational facilities and research institutes are using remote scientific packages to perform their detailed calculations. Similarly, mathematical modeling of business problems is used to project trends and aid in the decision-making process.
- Brokerage houses have built a very powerful, dedicated message-switching system to provide real-time updates of stock market information. Similarly, they now use remote computer operations for billing purposes.
- Airline reservation systems, such as United's worldwide data network, provide, via remote access, such things as:
  a. Flight information.
  b. Centralized inventory control of seats.
  c. Control of space allocation of off-line offices.
  d. Mechanization of passenger files.
  e. Wait listings, reconfirmation, and checking ticket time limits.
  f. Provision of special facilities for the passenger.
  g. Message switching.

h. Passenger check-in at airports.

i. Load and trim calculations.

j. Cargo reservations, etc. (see Applications).

Computer usage   As indicated above, computer usage is a direct function of computer availability. With the advent of the personal computer, which has been more and more sophisticated, the next step, interconnecting personal computers and mainframes, will promote more usage in more applications.

Data projections   Gross statistics can be given for major groups, based upon various projection techniques. However, we should realize that a projection is usually made by extending a trend, relating to another industry, or by curve fitting. Or, the consumer demand curve may be established by first reviewing with the potential customers the new, future products that may be available, and then showing potential benefits in order to determine their acceptance, changes, or nonacceptance. Since the data transport industry is still relatively new and not independent, but has a cross-elasticity relationship to the computer industry, transmission media, terminal markets, and programming language advancements, it's too early to project by traditional macroeconomic models, but may be approached by micro-economic techniques such as demand forecasting.

Projecting the volume of data calls and transactions for the 2000 + years has become quite complex. Computer usage breeds computer dependency, which breeds more computer usage. This increase in usage and sharing of costs brings down costs which, increases user demand. Usage increases as life habits change, where hard copy replaces oral or handwritten communications, as doctors, druggists, and police officers switch from oral conversations to data conversations, thereby decreasing the volume of voice communications and increasing the number of data transactions. So, let's take a closer look at what services are needed for the turn of the century.

Global information services

As technology experts explore the what and the how of various technologies, marketing researchers and planners address the where, when, why and for whom aspects of these forthcoming technical possibilities, as they use preprogram, preproject planning processes to translate these technical concepts to realistic applications. As part of this planning process of better understanding the customer, a series of trials need to be conducted to determine customers' needs and preferences. In so doing, a series of evolving technologies can be applied to specific applications, from medicine, to finance, to manufacturing, to real estate, to the executive boardroom. As we look at the future and what technology has to offer, we should view it as an overlay of narrowband, wideband, and broadband opportunities. With this in mind, we can look at the marketplace in terms of these three segmented groups.

Narrowband

As we use the narrowband technology of ISDN, we see that it indeed provides the opportunity to offer a public data network to our traditionally voice-only world. In the past, we've encouraged data users to interconnect their personal computers (PCs) using analog modems to transmit their information at low speed rates of 2400 to 9600 bits per second. By not offering other rates and the ability to switch data in a packetized mode, we've encouraged them to formulate their own internal LANs and interconnect them through point-to-point, leased-line, special service facilities. With the advent of T1, at 1.54M b/s, this became, in the mid '80s, the tool for interconnecting LAN arenas. However, in truth, though internal LANs operated at 2-10M b/s, the devices were, in essence, operating at much lower speed rates, using these higher speed backbones to ensure that there was sufficient capacity to move this information appropriately in real time.

As we look at the marketplace, there's indeed a need for narrowband ISDN to interconnect PCs from the home to the office, doctors to drugstores, lawyers to databases, both internally and externally. Similarly, there's a need for expanded voice services, such as CLASS, based upon using the calling-party number at the called-party location, enabling one to have the phone identify the person calling, even by name, or to perform special, selected-party operations. As we look at alarm monitoring, energy management, and remote meter reading, there are numerous low-speed data sensing and polling applications that can be transported over the narrowband facilities across the 18 major business sectors, from home to education. Hence, there's an opportunity for using narrowband ISDN over the existing RBOC copper plant. This was adequately demonstrated in early trials for moving X-rays to the home at 64K b/s and 128K b/s. Subsequent trials in the '90s have demonstrated these information-handling capabilities to the medical industry, as well as to others, such as real estate home listing.

Wideband

Another aspect of data internetworking is to interlink the various LANs together using a public packet-switched frame relay (connection-oriented) and SMDS (initially connectionless) based network. Here we can use various switching suppliers' capabilities to effectively achieve virtual private networks, as users determine their own configurations on a dedicated or switched basis. Similarly, T1 hubbing and T3 hubbing, as well as fractional T1, fractional ISDN primary rate, and perhaps, even fractional T3, can be demonstrated using digital cross-connects to enable customers' dynamic access to variable channel capacity and the automation of mainframe connections. We've also seen the need for establishing switching nodes closer to the users, from both an economic-sharing and a service-pricing point of view. Hence, these wideband switch points, which will evolve to broadband switch points, become, in effect, front-ends to the future higher-bandwidth switching network. Therefore, the objective of joint provider-supplier-user planning/trial activities is to help clarify these technical internetworking needs for this first phase of broadband services.

Broadband

Broadband services are just being identified and understood, as we pursue applications such as videophone, high definition TV, image transfer, supercomputer-to-supercomputer connectivity, as well as "high information content" home and business information generation, storage, access, manipulation, and presentation uses. This technology requires new fiber deployment plans and new switching capability. We must determine the most appropriate ATM or STM technology for variable-bit-rate (VBR) and continuous-bit-rate (CBR) traffic mixes. With this in mind, numerous videophone trials have been launched in the early '90s, using eye-to-eye terminal capabilities at various speeds. Some were initially switched at analog 56K b/s; other programs are progressing from 64Kb/s to 384K to 1.5M b/s, 45M b/s, and eventually to 155M b/s. (Realizing, of course, that at the higher 155M b/s rate, the information may actually be used at some lower rate.) (See FIG. A-4.)

As we move from the telephone (voice only) to the videophone (including data-handling capabilities), we need to understand the specific requirements for the various types of data, image, text, and video multimedia applications. This is a new industry; it's time to use today's technology to meet immediate needs and encourage user acceptance and growth. It's time to add a feature on a feature with a twist here and a turn there to tailor our offerings to their specific applications. It's a time for suppliers, providers and users to work together to identify the technology that they'll need in the future, by establishing a realistic family of equipment specifications for the switching, transport, and CPE systems that providers will need in order to meet their customers' needs in a timely manner. (See FIG. A-5.)

The following tables provide an exhausting and inclusive review of the technological possibilities and market opportunities of the narrowband, wideband, broadband—voice, data, text, image, and video offerings for the generators and users of information in the Global Information Society, in the next millennium—The Information Millennium. (See FIG. A-6.)

> "'The time has come,' the Walrus said,
> 'To talk of many things:
> Of shoes—and ships—and sealing wax—
> Of cabbages—and kings—
> And why the sea is boiling hot—
> And whether pigs have wings.'"
>
> **Carroll**

Fig. A-4. Narrowband-wideband-broadband features and services.

Narrowband – Copper
64K b/s

Voice Messaging, CLASS, Selected Messaging, Ckt Switching, Packet Switching, Protocol Conversion, Error Rate Control, Alternate Routing, Priority Messaging, E-Mail, Delayed Delivery, Voice/Text, Data Base Access, Encryption, Audit Trials, Broadcast, Polling . . .

Data Networks
Stock Exchange Networks
Police/FBI Networks
State Agency Networks
Medical/Insurance Networks
Auto Parts Networks
Auto License Networks
Inventory Control Networks

Financial
Legal
Small Business
Residential

Wideband – Copper – Fiber
(64K)–1.5M b/s

Bandwidth Management, Wide Area Network, Dynamic Bandwidth, Private to Public Internetworking, ISDN-Non ISDN Internetworking, POP Access, Channel Switching, . . . Survivable Private/Public Information Internetworking

X-Rays, Patient Records
CAD/CAM Networks
Graphic Display Networks
Computer to Computer Networks
1.5M b/s Picturephone Networks
Video Conference Networks
Wide Area Networks

Medical
State
Manufacturing
Securities

Broadband – Fiber
N (50M b/s)

Multiples of 50M b/s Switched Transport Video Conferencing, Picturephone, HDTV, Computer to Computer, High Speed Data Transfer, SMDS, FDDI, Frame Relay, Call Relay (ATM), & Broadband Information Transfer, Storage, Access, & Presentation

High Resolution Picturephone
Entertainment
Media Events
Education Video
Medical Imaging
High Definition CAD/CAM
High Definition TV
Computer to Computer
Data Base Manipulation
Visual Presentations

Education
Entertainment
Large Business
Residential

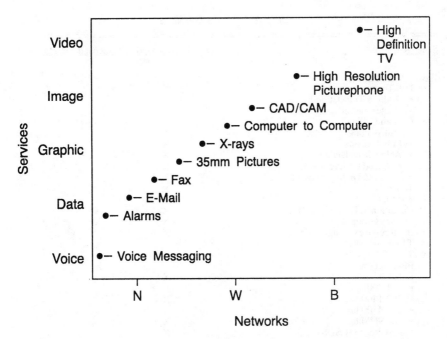

Fig. A-5. *Customer needs/services/networks.*

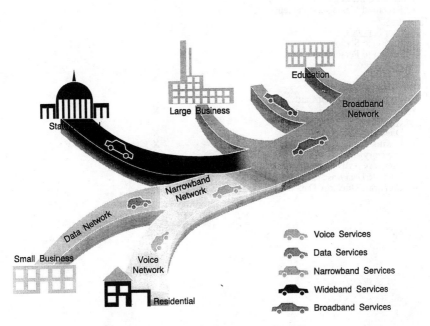

Fig. A-6. *Future pathways to the information marketplace.*

Table A-1. Voice Services

SERVICES	Narrowband	Wideband	Broadband
Basic Voice Network			
• Dial Tone	X		
• Address Translation	X		
• Route Control	X		
• Custom Calling	X		
• Call Waiting	X		
• Call Transfer	X		
• Abbreviated Dialing	X		
• Speed Calling (CPE)	X		
• Speed Dialing Network	X		
• Touch Tone	X		
• 2nd Line	X		
• Conditioned Line (C1, C2)	X	X	
• Feature Group A, B, C, D	X	X	
• fx - foreign exchange line	X		
• T1 - 1.544 Mb/s		X	
• T3 - 45 Mb/s		X	
• PBX Access		X	
• Data Set	X		
• 101 300 b/s	X		
• 102 1400 b/s	X		
• 103 4800 b/s	X		
• 104 9600 b/s	X		
• 56K switched (CSDS)	X		
• Number Movement	X		
• Fast Provisioning	X		
• Weather (Audio Tex)	X		
• Time (Audio Tex)	X		
• Credit Card Call Billing (Telco)	X		
• Recorded Messages (Intercept)	X		
Operators			
Operator Services			
• Station to Station	X		
• Person to Person	X		
• Collect	X		
• Third Party	X		
• Calling Card			
• Private Label Credit Card	X		
• Multiple Language	X		
Directory Services			
• Directory Assistance	X		
• Locator Service	X		
• Directory Assistance Call Completion	X		
• Directory Assistance Virtual Private Network	X		
• Multiple Data Base Access	X		
• Corporate Directory Data Bases	X		

Table A-1. Continued

SERVICES	Narrowband	Wideband	Broadband
Conference Calling			
• Auto Dial	X		
• Conference Set Up (Intra)	X		
• Music on Hold	X		
Customized Services			
• International Assistance	X		
• Dual Party Relay Services for Hearing Impaired	X		
• Private Network Services	X		
• Order Fulfillment	X		
Telemarketing			
• Data Base Access (LEC)	X		
• Simultaneous Voice	X		
Enhanced Centers			
• Voice Messaging	X		
• Voice Recording	X		
• ISDN Access	X		
• Dual Line	X		
• Number Change	X		
• Data CKT Switching	X		
• ACD Automatic Call Distribution	X		
• Auto Attendant	X		
• Key ISDN	X		
• Key/PBX	X		
• ECD (Electronic Call Distributor)	X		
• Extended Customer Control - bandwidth, routing	X		
• Co LANS		X	
• Data Access		X	
• LAN Access		X	
• Integrated Voice/Data Access	X		
• Access to CSDS "56"K Net	X		
(Circuit Switched Digital System)	X		
• Network Modem Pooling	X		
• Private Virtual Networking	X	X	
Enhanced 911			
• Call Party Control	X		
• Emergency Ring Back	X		
• Forced Disconnect	X		
• Idle Circuit Tone Applicator	X		
• Route Diversity	X		
• Calling Party Switch Hook Status Indicator	X		
• ANI	X		
• Selective Routing	X		
• ACD	X		
• Call Add On	X		
• Computer Answered Dispatch	X		
• Automatic Location Information (ALI)	X		
• End User Verification of ANI/ALI	X		

SERVICES	Narrowband	Wideband	Broadband
Voice Mail			
• Special Delivery	X		
• Special Handling	X		
• Dial Dictation	X		
• Junk Mail Blocked (only certain calls accepted)	X		
• Equal Access to IXC	X		
• Voice Response	X		
Open Network Architecture			
• Basic Serving Arrangement (BSA)	X	X	X
• Basic Service Element (BSE)	X	X	X
• Complementary Network Service (CNS)	X	X	X
900 Access Services			
• 976 Dial a Porn	X		
• 976 (Blocked) - Screening	X		
- Access Code	X		
• Advertising (Menu)	X		
• Special Orders (Menu)	X		
• Personal Messages (Menu)	X		
• Audio Tex Messages (Menu)	X		
• Gab Lines	X		
• Customer Service Hotline	X		
• 900 Service Bureaus (Menu Only)	X		
• Entertainment 68%	X		
• Polling 10%	X		
• Live Applications 15%	X		
• Promotions 3%	X		
ANI/SS7			
• "D" Channel Signalling Services	X		
• "D" Channel Packet Network	X		
Class			
• Calling Number Delivery	X		
• Call Identity	X		
• By Number	X		
• By Name	X		
• By Recorder	X		
• Call Block (selected)	X		
• Call Transfer (selected)	X		
• To Police	X		
• To Cellular Phone	X		
• To Special Message Recording	X		
• To Voice Mail	X		
• To Voice/Tex	X		
• Automatic Recall	X		
• Repeat Dial	X		
• Calling Number Suppression (Optional)	X		
• Voice Response	X		
• Selective Call Rejection	X		
• Selective Call Forwarding	X		
• Customer Originated Trace	X		
• Calling Number Displayed	X		
• Distinctive Ringing	X		

SERVICES	Narrowband	Wideband	Broadband
Class (continued)			
• Call Screening	X		
• Call Priority	X		
• Override	X		
• Reroute	X		
• 800 "Toll Free" Services (Within LATA)	X		
Intelligent Network Service			
• Local Area Signalling System	X		
• Real Time Authorization Code Verification	X		
• Calling Number ID	X		
Enhanced Pay Phones			
• Speech Synthesis	X		
• Calling Card Reading	X		
• Call Screening	X		
• Call Redirect	X		
• Data Communications	X		
• Alpha/Numeric Display	X		
Personal Services With Computer Interaction			
• Speech Recognition	X		
• Speaker Dependent	X		
• Speaker Independent	X		
• User Interaction	X		
• Talking Yellow Pages	X		
• News Announcements	X		
• Sports News	X		
• Stock Market Information	X		
• Name of Person Calling	X		
• Computer Asks What To Do Next	X		
• Computer Provide Options To Users	X		
• Special Messaging, Call Forwarding	X		
ISDN CPE (Better, Faster, Smaller, Smarter,	X		
High Function, High Performance CPE)			
High Tech CPE Phone Features			
• Voice to Text	X		
• Auto Answer	X		
• Programmable Ring	X		
• Station BLF	X		
• Adjustable Volume	X		
• Computer/Message Display			
• Messaging	X		
• Message Logging	X		
• Text Messaging	X		
• Call Cost (Cost Accounting)	X		
• Call Duration (Time Accounting)	X		
• Silent Communications	X		
• Touch Screen	X	X	X
• Touch Tone	X		
• Auto Call Back	X		
• Call Forwarding	X		
• Calling Name	X		
• Calling Number	X		

SERVICES	Narrowband	Wideband	Broadband
High Tech CPE Phone Features (continued)			
• Call Pickup Service	X		
• Message Warranty Service	X		
• Priority Call	X		
• Voice Command Recognition	X		
• Desk-to-Desk Messaging	X		
• Multiple Simultaneous Conversations	X		
• Hands-free Calling	X		
• Speed Calling Directory List	X		
• One Number Services	X		
• Flexible Virtual Private Networking	X	X	X
• Personal Identification Number (PIN)	X		
• Customer Reconfiguration	X		
• Traffic Analysis	X		
• Service Restriction	X		
• Routing Controls	X		
• Call History Date	X		
• 800 (LATA)	X		
• Calling Card	X		
• Credit Card	X		
• Area Wide Centrex	X		
• Interactive 800	X		
• Dialed Number Service (700/900, tele, rotary network ACD, flexible charging service)	X X		
• Network Anywhere Call Pickup	X		
• Personal Number	X		
• Reserved Conference Call	X		
• Security and Screening	X		
• Network Access	X		
• Computer Access	X		
• Messaging Services	X		
• Wake-Up Call	X		
• Dynamic Multimedia Teleconferencing	X		
• DTMF (Dialtone Multi Frequency) Signalling Services	X		
Network Management			
• Routing	X	X	X
• Reroute	X	X	X
• Access	X	X	X
• Blockage	X	X	X
• Layers	X	X	X
• Private	X	X	X
• Public	X	X	X
• CPE			
• Real Time Disaster Recovery	X	X	X
• Network Status	X	X	X
• Time Variant Service Options	X	X	X
• Time Variant Bandwidth	X	X	X
• Time Variant Circuit Connections	X	X	X
• Permanent Cross Connections	X	X	X
• Nail Up	X	X	X

Table A-1. Continued

SERVICES	Narrowband	Wideband	Broadband
<u>Customer Services</u>			
• Controlled Reconfiguration	X	X	X
• Bandwidth on Demand	X	X	X
• Dynamic Bandwidth	X	X	X
• Alternative Routing	X	X	X
• IXC Preference	X	X	X
• Private Network Backup	X	X	X
• Private Network Access	X	X	X
<u>Service Creation</u>			
• Fast Provisioning	X	X	X
• Voice Data Testing	X		
• Specialized Billing	X		
• Alternate Billing	X		
• Credit Card	X		
<u>Enhanced Centrex</u>			
• Voice Mail	X		
<u>ISDN High Fidelity Audio</u>			
• High Fidelity Voice	X		
• 7 Khz Audio Stereo not 3.1 Khz	X		
• Better Speech Recognition	X		
• Clarity & Quality Voice	X		
• Improved Speech to Tex	X		
• Bit Error Rate 10^7 - 10^{11}	X		
<u>Shared Tenant Services</u>			
• Tenant Portioning	X		
• Automatic Route Selection	X		
• Advanced Network Features	X		
• Modem Pool	X		
• Protocol Conversions	X		
• Integrated Voice & Data	X		
• Text Processing	X		
• Video Conferencing	X		
• Tenant-to-Tenant Services	X		
• Energy Management & Control Systems (EMCS) (CPE)	X		
• Paging Service	X		
• Telex	X		
• Facsimile	X		
• Audio Conferencing	X		
• Video Tex (Gateway)	X		
• Electronic Directory	X		
• Archival Storage	X		
• Voice Mail	X		
• Access to Remote Data Bases	X		
<u>PBX Automatic Attendant Services</u>			
• Enhancement, not Replacement (CPE)	X		
• Customized Menus	X		
• Customized Routing Path to Specific Departments	X		
• Voice Mail	X		
• Dial Dictation	X		
• Data Access	X		

SERVICES	Narrowband	Wideband	Broadband
PBX Automatic Attendant Services (continued)			
• Outgoing Network Access	X		
• PBX Inward Dialing	X		
Access to Available Intelligent Network Platforms (INPs)			
• Message Arch Services (Voice)	X		
• Software Portability	X		
• Flexible End User Access	X		
• Applications Programming	X		
• Capability for New Services	X	X	X
• User Information Platform Access	X	X	X
• Network Information Platform Access	X	X	X
• Information Services Platform Access	X	X	X
• Access to Enhanced Service Provider	X	X	X
• LAN Network Connectivity Service (LCS)	X	X	X
• Shopping Bag Gateway to Data Bases	X		
• Dialup Data	X		
• Voice Access	X		
Cellular Phone			
• Bill Calling/Response			
• Call Transfer			
• Data Messages			
Personal Communications Network			
• Personal Phone			
• Personal Identity Number (PIN)			
• Telepoint			
• PTN (Private Telephone Number)			
• Mutiparty/Single Party			
• 450 Mhz with 650 Khz bandwidth for 570 Users			
• 150 Mhz with 650 Khz bandwidth for 570+ Users			
• 850 Mhz with 650 Khz bandwidth for 570++ Users			

Table A-2. Data Services

SERVICES	Narrowband	Wideband	Broadband
Data Transport			
• Dialup Modems (Analog)	X		
300 b/s V.21	X		
1200 b/s V.22	X		
2400 b/s V.22BS	X		
4800 b/s V.27 Ter	X		
9600 b/s V.32	X		

Table A-2. Continued

SERVICES	Narrowband	Wideband	Broadband
Data Transport (continued			
• Leased Line Modems (Analog)	X		
1.2Kb/s	X		
4.8Kb/s	X		
9.6Kb/s	X		
14.1Kb/s	X		
19.2Kb/s	X		
• 56Kb/s CDCS	X		
• Digipac	X		
• Dialup Access	X		
Gateways			
• Multi Tier/ - $10/mo to $80/mo Tiers	X		
(Menu only within LATA)	X		
• Browse within a Tier	X		
• Voice Messages	X		
• Data Messages (E-Mail)	X		
• Data Bases	X		
• Dialup Voice Data	X		
• Direct Data	X		
ISDN Basic			
• "D" Channel Signalling (Transport/Only)	X		
• Alarms	X		
• Energy Management	X		
• Meter Reading	X		
• Terminal to Terminal	X		
• Terminal to Network	X		
• "B" Channel "CKT" Service (64Kb/s)	X		
• Permanent	X		
• Dial Up	X		
• "B" Channel "PKT" Service (64Kb/s)	X		
• "D" Channel "PKT" Service (9.6Kb/s)	X		
ISDN Primary			
• DS1 Switched		X	
• DS1 Non-Switched		X	
• N x DSO (n x 64K)		X	
• H0 384K		X	
• H11 768K		X	
• H12 1.536K		X	
• "D" Channel Services		X	
Broadband			
• PBX Direct	X		
• Inward Dialing	X		
• ANI	X		
• Billing	X		
• Fractional T1	X		
• 384K	X		
• N (64K)	X		

SERVICES	Narrowband	Wideband	Broadband
Exec Workstations			
• Auto Dial	X	X	X
• Auto Login	X	X	X
• Windowing	X	X	X
• Multiple Screens	X	X	X
• Multiple Sessions	X	X	X
• Multiple Protocols	X	X	X
• Programmable Buttons	X	X	X
• Terminal Emulation	X	X	X
• Distribute Processing	X	X	X
• Parallel Processing	X	X	X
Text Messaging			
• E-Mail "Secure Domains"	X		
• X.400 to Link E-Mail Islands	X		
• Video Tex (Gateway)		X	X
• X.400 Electronic File Transfer, Messaging	X		
• X.500 Electronic Directory	X		
• Computer Security	X	X	X
• Junk Data Mail Filter - ID, Audit Trails, Access, Verification, Log History, Terminal Verification	X		
• Word Processing	X	X	
• Local Networks		X	X
• Eithernet 802.3, Token Bus 802.4 & Token Ring 802.5		X	X
Data Applications			
• TeleCommute, TeleDoctoring, TeleMarketing, Tele . . .	X	X	X
Data Interfaces			
• TCD/IP (Transmission Control Protocol/Internal Protocol)		X	X
• MANS (802.6) Metropolitan Area Network		X	X
• FDDI		X	X
• Connectionless (SMDS)		X	X
• API - Applications Program Interface (Transport only)	X	X	X
• Example: Hayes ISDN for PC	X		
• TA = Terminal Adaptor	X	X	X
• X.400 and X.500 Packages for terminals to access	X		
• X.400/500 vendors	X		
• Interactive Data Bases	X	X	X
Broadband			
• VANS - Connection Oriented (LATA/Transport only)		X	X
• Frame Relay		X	
• LAN to WAN Interface ANSI STD		X	
• Variable Length Packets		X	
• Interface to CELL Relay Systems		X	X
• Wideband Packet Switching		X	
• Fast Packet CCITT 1.121		X	X
• Cell Relay		X	X
• 48 Cells + 5 Header Cells		X	X
Facsimiles			
• Group 3 (2.4K, 9.6K), Group 4 (56/64K)		X	
• Group 5		X	
• S/F	X		
• Delayed Delivery	X		

SERVICES	Narrowband	Wideband	Broadband
Facsimiles (continued)			
• Protocol Conversion	X		
• Repackaging	X		
• Closed Service Groups	X		
• Gateways	X		
• Broadcast	X		
• Polling	X		
Data Handling: Automatic Teller Machine			
• Transactions	X		
• Remote Banking	X		
• File Transfer	X		
• File Access	X		
• Polling	X		
• Broadcast	X		
• Priority Messages	X		
• Delayed Delivery	X		
• Foreign Language Convertors	X		
• Special Handling	X		
• Printer Backup (Off Line)	X		
• Gateway to Data Base	X		
• Menu	X		
• 3rd Party Software	X		
• Enhanced Service Provider	X		
• List Processing	X		
• Imaging	X		
• Protocol Conversion	X		
• Code Conversion	X		
Wide Area Network (Per LATA)			
• 64K	X		
• N (64K)		X	
• PVN (Private Virtual Network)	X	X	
• T1 - 1.544 Mb/s		X	
• Fractional T1, F-ISDN		X	
• T3 - 45M b/s		X	
• Wide Area Network Centrex		X	
• Co-LAN		X	
• LAN to Public Network	X	X	
• Address	X	X	
• Routing	X	X	
• Network Management	X	X	
• Bridging	X	X	
• Gateway	X	X	
• Error Detection/Correction	X	X	
• High Speed Delivery	X	X	
• Packaging	X	X	
• Packet Sequencing	X	X	
• Packet S/F	X	X	
• Packet Delayed Delivery	X	X	

Table A-2. Continued

SERVICES	Narrowband	Wideband	Broadband

Wide Area Network (continued)
- "D" Channel Service (SS7)   X   X
- Terminal to Terminal   X   X
- Terminal to Computer   X   X
- Computer to Computer   X   X
- Terminal to Network   X   X
- Network to Network   X   X

Storage
- Optical Disks
  - Read Only
  - Write Once
  - Erasable
- ROM - Read Only Memory
- Compact Disk - ROM (CD-ROM)
- WORM - Write Once Read Many Times
  - 5 1/4" - 350Mb/s
  - 12" - 15 gigabytes

Data Pay Phone
- Portable Computer
- Basic ISDN
- Primary ISDN

Data Messaging
- PC to PC Network
- V.120 "B" Channel Data, API
- X.400 Gateway Links Multiple LAN based E-Mail System or X.400 inter portability between VANS for different 'body types', for example, IBM Document Interexchange Arch, /Document Content Arch, Distributed Office Support System (DIA/DCA) but not DISOSS - X.409 format for listing type of contact in document. On Time Protocol Adaption Language (OPAL) - would allow translating in C, PASCAL or COBOL

Information Gateways
- Interface to Data Bases
- Customized Operations
- Electronic Meeting Phone
- Broadband Selection of Services
- Shopping Bag Gateways
- Kiosk Billing
  - Length of time - no service change within Tier

Compression Techniques
- ADPCM (Adaptive Quantization Pulse Code Modulation)

Computer Power
- Typewriters to Word Processors to Personal Computers to Mainframes to Distributed Processing (Transport) Reduced Instruction Set Computers (RISC); for example, 32 bit 11-12 MIPS Fault Tolerant Unix Sy.5 Automatic Teller Machines (ATMs)

SERVICES	Narrowband	Wideband	Broadband

Security
- User verification using biometric techniques
  - Voice
  - Fingerprint
  - Eye Retina
  - Speaker Verification
  - Expect system for tracking usage

Portable Electronic Terminals
- Debit Card (Sales terminals - cannot operate the system)
- Smart Card
  - PIN - Personal Identification Number
  - Medical History, Social Security Number
  - Insurance Policy
  - Patient Records
  - Eating History
  - Medical Alerts
  - EEPRON - Electronically Programmable Read Only Memory Microprocessor
    - 1 chip = 8K memory
    - 2 chips - 64K memory

Laser Card (Off Line)
- Optical Storage Cards $2-3 (Sale not operate)
- Uni-card $20
  - Keyboard
  - Memory
  - Display
- Ulti-cord $20
  - Computer
- Magnetic Strip Card $1

Data Services	Narrowband	Wideband	Broadband
• Store & Forward Features	X		
• Broadcast "CC" Carbon Copies	X		
• Electronic "Yellow" Pages	X	X	X
• Electronic "White" Pages	X		
• Pay Bills	X		
• Transfer Money	X		
• Review Accounts	X		
• Order Theater Tickets	X		
• Make Travel Reservations	X		
• Calculate Taxes	X		
• Motor Vehicle Registration	X		
• Home Shopping	X	X	X
• Home Banking	X		
• Remote Medical Exams	X	X	
Data Transport (Old)			
• Telex 300 b/s	X		
• Telex 1200 b/s	X		
• TWX - 110 bit/sec ASCII	X		

Table A-2. Continued

SERVICES	Narrowband	Wideband	Broadband
E-Mail			
• Message	X		
• Confirmation	X		
• Forwarding	X		
• Annotation	X		
• Editing	X		
• Adding	X		
• Inserting	X		
• Distribution Lists	X		
• Recording Messages	X		
• Receiving Messages	X		
• Saving Messages	X		
• Forwarding Messages	X		
• Replying to Messages	X		
• Confirmation to Reception	X		
• Reviewing before Sending	X		
• Scanning Messages	X		
• Selectively Ordering Messages	X		
PC - Service Bureau			
• Telex	X		
• Facsimile	X		
• Video Tex	X	X	
• Electronic Mail	X		
PC Host			
• Micro - Main Frame			
• Distributed Processing			
• Cooperative Processing			
• Shared Processing			
• File Transfer			
• Front End Processing			
Data Base Management			
• SQL - Structured Queuing Language Data Base Service			

Table A-3. Video Services

SERVICES	Narrowband	Wideband	Broadband
Video/Imaging Services			
• Picturephone 56K, 64K, 364K, 1.5M, 45M, 140M	X	X	X
• Still Picture (64K)	X		
• X-Ray (128K, 1.5Mb/s)	X	X	
• Picture Quality Image (1M bits) at 1.5M b/s		X	
• Video Conferencing 1.5M b/s, 45M b/s		X	X
• Advanced TV 100M b/s			X
• High Definition TV 140M b/s, 620M b/s			X

Table A-3. Continued

SERVICES	Narrowband	Wideband	Broadband
<u>Hyper Text</u>			
• Process			
• Ways to store information in discreet chunks			
• Sentences, Paragraphs, Pictures			
• Ways to view information two or more ideas at a time			
• Ways to form connections to items			
• Network nodes data bases			
• Process			
• Links between nodes			
• Annotational			
• Composition process			
• Gateway access			
<u>Network Management</u>			
• Gateway to internal OSS		X	X
• Circuit Status Data		X	X
• Service Orders		X	X
• Traffic Reports		X	X
• Configurations		X	X
• Routing		X	X
• Performance Management		X	X
• Testing		X	X
• Billing Information		X	X
<u>Video-DS3</u>			
• Educational TV			X
• Remote Classview			X
<u>Video Programs</u>			
• Access to All Channels			X
• Dialup (4)			X
• Browse			X
• Picture in a Picture			X
• HDTV			X
• ATM			X
• STAR			X
• Broadcast			X
• 3D Home Video			X
<u>Medical Imaging</u>			
• PACS - Picture Archiving and Communication System	X		X
• Remote CatScan (RCS)		X	X
• Magnetic Resinous Imagery (MRI)		X	X
• Computer Ordered Tomography (CAT)		X	X
Video Multimedia Conferencing		X	X
• Simultaneous Text, Graphics, Voice and Video		X	X
• Super Computer Interactive distributed file systems		X	X
• Browse		X	X
• Changeability		X	X
<u>Video Transport</u>			
• HDTV			X
• Improved Definition 30-40%			X
• EDTV/SVHS/ACTV			X
• Enhanced Definition			X
• Advanced/Compatible (25% better)			X

Table A-3. Continued

SERVICES	Narrowband	Wideband	Broadband
<u>Video Transport</u> (continued)			
• Video Services		X	X
• Shopping		X	X
• Finance		X	X
• Travel		X	X
• Entertainment		X	X
• Education		X	X
• Information		X	X
• Video Library			X
• Broadband Video			X
• Video Data Base			X
• OCR - Optical Character Recognition		X	X
• Still Video STDS		X	X
• Computer Aided Design (CAD)		X	X
• Computer Aided Manufacturing (CAM)		X	X
• Computer Aided Education (CAE)		X	X
• Full Motion Sports Events			X
• Electronic Photographics Storage		X	X
• Interactive TV (2 Way)			X
• Corporate (Video Conferencing) TV		X	X
• Video Advertising		X	X
• Video Security		X	X
• FTTP/FTTH			X
• Fiber to the Pedestal			X
• Fiber to the Home			X
• 600M b/s-2.46 b/s transport			X
• 4-150M b/s			X
• SMDS (Public)		X	
• Connectionless		X	
• LAN to LAN		X	
• SONET (40M b/s)		X	
• FDDI (Private)		X	
• 100M b/s		X	
• LAN to LAN		X	
• Connection Oriented		X	
<u>Passive Vrs Active Optics</u>			
• Broadcast Service, HDTV (Transport)			X
• Multiplexing/Splitting			X
• Passive diffraction for different wavelengths			X
• Active fiber loop optical transceivers			X
<u>Fiber Topology</u>			
• New Topology		X	X
• Class 5		X	X
• Class 6		X	X
• FTTP		X	X
• Class 7		X	X
• Class 6		X	X
• DACS		X	X
• ATM		X	X
• SMDS		X	X
• Private to Public		X	X

SERVICES	Narrowband	Wideband	Broadband
Fiber Topology (continued)			
• TS1		X	X
• Optic Switching		X	X
• Multiple Base Units Access		X	X
• Public - POPs Access		X	X
• 6 to 6 Networking		X	X
• T1/FT1/F/ISDN		X	X
• Channel Switching		X	X
• Bandwidth on Demand		X	X
• Internetworking	X		
• ISDN	X		
• Non-ISDN	X		

"'But wait a bit,' the Oysters cried,
'Before we have our chat;
For some of us are out of breath,
And all of us are fat!'"

Carroll

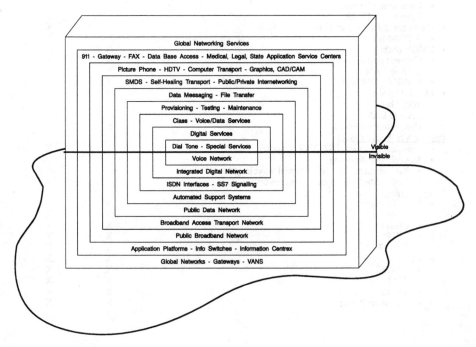

Fig. A-7. Narrowband-wideband-broadband information services.

Appendix B

Future issues & strategies

As we consider future issues and strategies, it should be quite apparent that several key aspects of information movement and management (IM&M) are very interrelated; we can't address applications without having a solid, supportive infrastructure, from which information services are deployed. However, some private users and providers might still believe that a public network infrastructure is not needed. They might believe that many, if not all, future services can be supported by private offerings. Hence, it's important to more clearly review several specific applications and their affected industries to better understand and differentiate what type of networks are needed, both private and public, to more fully meet the complete community of interest of each application. With this in mind, let's take a more detailed look at several applications and services of the forthcoming information infrastructure, as we consider future issues and strategies.

Applications

In order to be market driven, there's considerable interest in first identifying user needs, and then defining the right technical solutions to meet these needs. This form of analysis leads to reviewing the user's mode of operation in specific applications. From this, we'll see how specific forms of communication can provide specific services that assist the user in completing his or her specific tasks, while eliminating old ways and enabling new ones. So, let's proceed with a more detailed look at the health-care industry to more fully appreciate the new telecommunications opportunities of the '90s and beyond.

Health care

As noted earlier, in FIG. 3-7, there's a wide community of interest within the medical community that crosses business sectors and market segments, as it spans both large and small business, as well as residential, insurance, banking, education, and state and federal agencies. To appreciate this wide breadth and scope of health care, let's review this industry from the perspective of a rural General Practitioner (GP); for the most part, the rural attribute doesn't diminish the urban doctor's needs, but simply adds one more dimension to the study.

For the sake of mutual support, social interaction, and customer convenience, many doctors within rural (or urban) communities seem to gather in clusters—in a medical center. Here resides the GP, along with the dentist, and one or more specialists (the gynecologist, urologist, or even the "skin doctor"—dermatologist). The more rural the location, the fewer the number of specialists. In fact, in some very rural areas, the nurse replaces the GP.

So, let's begin by assuming that you're the local general practitioner (the "doc") in Vista Falls, CO, elevation 7,500 feet, wintertime population 3,000, as the skiers and the service providers triple the local summertime population of 1,000. Though this imaginary village is 50 miles from Denver, it functions as an extended suburb, using as many medical facilities as possible from the Denver area. However, when weather conditions dictate, the local medical center has to function as the local hospital—sometimes for an extended period, until storms subside and mountain passes are cleared. Two doctors, a secretary, three nurses, a midwife nurse, a dentist, and a part-time X-ray technician support the entire basin. Many of the residents consist of the highly educated, middle-aged group, who have moved out from the city to enjoy the country life, but continue in many diverse businesses that can be accommodated from one's home or by one or two road trips per month. The remaining group consists of local ranchers, the village store owners, the service providers, and those employed in the nearby large ski resort.

As you rise early to begin your daily tasks, it's important to understand with whom you need to communicate to best understand your communication needs, beginning with the patient, who may have come to the office, or called from home, or is in the Denver hospital. You may also need to review new medicines available from drug companies located across the country, send prescriptions to the local druggist, determine patient status in the distant hospital, review medical alerts from state and federal agencies, discuss X-rays with the Denver hospital or clinic radiologist, receive and analyze results from blood tests, or communicate analyses, X-rays, and test results to referral surgeons and consulting specialists.

These operational endeavors, for the most part, require narrowband data communications, as individual reports and analyses are transported here and there throughout the area and country. Internal patient records are distributed as needed externally, while admission records are selectively processed in moving the patient from the doctor's office to the specialist or the hospital. While wideband facilities will speed up and enhance video resolution and transmittal of X-rays, broadband networks will facilitate educational endeavors, as remote (or local) doctors attempt to maintain their education in new procedures, drugs, and research developments.

The key to all this information handling and sharing will be the availability of the narrowband, wideband, and broadband networks that enable the full movement and management of medical information. While the remote areas will have less-frequent usage, they will require access when and where they need it; on the other hand, the more urban communities will have higher concentrations of specialists, who will continually share and exchange increasing amounts of information—especially within a closed community, such as within a hospital, and to and from a nearby large medical center.

As data traffic increases and more and more users exchange information, it's necessary to increase both CPE and network transport capabilities. One technician can manually send X-rays to several remote doctors, but a more automated facility will be needed, such as the automated archiving of many patient records, so that numerous doctors can access these historical files and obtain online real-time test results from more sophisticated machines such as the "doughnut" Magnetic Resonance Imagery (MRI), which can generate as many as 100 X-rays to fully describe the patient's condition.

To accomplish this movement of more and more information to more and more users in real time, internal data transport local area networks have increased from 2 million bits per second, to 10, and on to 16 Mb/s. The increased rates are not necessary for higher terminal speeds, but the rates are required to handle the greater volume of terminals still generating lower-speed data. In time, however, as the user demands more and more imagery of higher and higher resolution to full-motion video educational films, the jump will occur from 16 Mb/s, or even 45 Mb/s, to 100 Mb/s and even several gigabits.

In this mode, the higher bandwidth traffic is needed for videophone, video conference centers, multimedia workstations, or knowledge stations that use video and three-dimensional imagery to show internal organs and conditions, as well as provide video displays for personal conversations. There will continue to be a need for narrowband facilities for moving patient data files and records here and there, as well as sending and receiving insurance claims and medical broadcast reports. Therefore, with new broadband Asynchronous Transfer Mode (ATM) and Synchronous Transfer Mode (STM), bit and byte, circuit, packet, and channel-switching technologies, it's quite possible to supersede LAN, WAN, and MAN technologies with publicly switched network offerings that handle individually switched information from autonomous terminals via direct fibers. In the interim, hybrid techniques for networking LANs and switching narrowband/wideband facilities will prevail until full broadband ISDN is available. The key to success for whatever bandwidth offering will be its ubiquitous availability and addressability within the user-group community of interest.

If needed, we could go on to review several other industries, such as a further case study of the banking industry, but the medical industry is quite sufficient and representative. Hence, it should suffice to skip to the issues and observations.

Issues and observations

We've seen the need for a narrowband data network to move patient records, prescriptions, test results, patient status, etc. It's true that the medical industry in particular is also moving to imagery and video, which can be handled by wideband initially, and eventually broadband, especially for full-motion, high-resolution video for real-time remote analyses or educational reviews. Therefore, it's essential to keep in mind the following key issues and observations.

To solve the tremendous congestion, overpopulation, and other social problems of our major cities, communication networks can enable remote communities to interrelate with urban services, as if they were a close suburb or actually a part of the urban scene. In this manner, distance becomes an unnecessary obstacle to human interaction. No longer do cities need to grow to megalopolises, as infolopolises are distributed throughout the region, connected by communication highways. As the world gets closer together, communities will turn to communications for the infrastructure backbone needed for economic survival in the new millennium; this requires a long-term commitment by each local community as it works with state and federal agencies to foster this change, to cap urban growth and establish new cities in remote places. How can urban leaders become cognizant of this opportunity and be encouraged to promote the long view? What role should federal, state, and local regulatory agencies play? What can the providers do to present and establish this vision of the future? How can affected industries, such as the medical community, participate? What can service providers do to encourage expenditures on the needed infrastructure—especially as it's quite evident that without an ubiquitous network, fewer large, corporate, internal facilities can be accessed and interrelated with small businesses or remote users?

Services

Services, services. As we proceed from voice services to data services provided by present and future data-networking techniques, it's important to realize that many of today's analog services are changing to digital services provided by n-ISDN and B-ISDN. So what are these new entities in light of existing data services, and what are their future issues and concerns?

Data services—data communications

Over the years, a variety of methods have been developed for "communicating" data information, that is, transmitting data over a transmission medium to a distant receiver, thereby achieving "data communications." The data is generally encoded in the binary form of "ones" and "zeros," usually represented by the presence or absence (on-off) of a signal or voltage condition. The data terminal equipment (DTE) is the terminal, printer, or computer that transmits or receives the information, while the data communications equipment (DCE) multiplexes, modulates or demodulates (modem), or concentrates the information over the medium. Information is transmitted from point to point or multipoint (where several DTE's share the same medium). There are several protocols by which terminals talk to each other to ensure that they exchange the information; there are also several interface standards, such as RS-232-C, by which physical and logical connections are achieved. Information can be sent over two-wire or four-wire systems in the *simplex* (data flows in one direction), *half-duplex* (both directions over the same path), or *full-duplex* (separate each way over four wires or split channels) form.

In considering previous data-handling methods for transporting bits and bytes, here are some interesting tidbits. Large central processing units (CPUs), sometimes called mainframes or host computers, become more and more useful as their DTEs become more interactive or conversational. Usually these terminals send information over low-speed modems in the asynchronous mode, using frequency shift keying techniques that send individual characters with a header and a tail. Some terminals are polled for information; others, such as smart terminals, having limited processing capabilities, or intelligent terminals, that process programs and send information in high-speed blocks synchronously to and from the mainframe. Personal computers have now proliferated, where they buffer (store) their results locally and interexchange results, processed data, or programs. They have become standalones or nice front-end processors for supercomputers. Thus, the more these devices (DTEs, CPEs, and PCs) are networked, the more efficient and effective they are in providing solutions for today's distributed applications. These distributed networks send information in analog form at 300, 1200, 2400, 4800, 9600, 14,400 and 19,200 bit rates. In earlier terminology, bit rates and baud rates were both referenced. Since they're the same at 300 bps (300 baud), the terms are interchangeable. But baud rate can't exceed the bandwidth of the line—for example, 2400 for the phone line, where, through coding techniques, more and more bits can be sent. For example, when 9600 bits/sec phase shift modulators are operating at 2400 baud (4 bits per baud). Hence, it has become less confusing to simply state bit rates. Similarly, coding standards have progressed from Baudot operating at a 5-level code, capable of generating 58 different characters; to ASCII, a 7-bit code; to EBCDIC (Extended Binary Coded Decimal Interexchange Code), an 8-bit code developed by IBM, capable of 256 characters. This latter code sets the stage for the 8-bit byte, where multiples of it achieve the 8-, 16- and 32-bit computer instruction or database data bank memory.

For error correction and detection techniques, bit and bytes were added respectively in order to form vertical or horizontal parity, enabling errors to be detected using cyclic redundancy checks (CPC), where vertical parity checks the character and the horizontal parity checks the block in terms of even or odd counts. As errors are checked and detected, the protocol of ACKs (acknowledgements) or NAKs (negative acknowledgements) ensure retransmission until the proper delivery is achieved.

In the early analog frequency baud transmission world, narrowband was subvoice or less than 300 Hertz (Hz), voice-grade baud 300-3000 Hz, while wideband or broadband required bandwidth considerably more than 2500 Hz, enabling rates up to 64,000 bps or better. Local area networks were established using metallic leased lines or private coaxial cables. Here, baseband facilities carried information to a distance of a mile or two, while broadband could accommodate many channels for greater distances of approximately 10 miles or so.

Analog modulation techniques used by modems included frequency modulation, noted earlier as frequency shift key modulation (FSK), where the frequency is varied while the height or amplitude of the waves is kept constant. While FSK is limited to low speed as it sends only one bit per baud, amplitude modulation is similarly limited, as it varies the height or amplitude at rates between 300 and 2400 bps. Phase modulation uses phase shifts of 180 degrees and can transmit up to 3 bits per baud. Quadrature amplitude modulation (QAM) combines both amplitude changes and phase shifts to achieve 4800 and 9600 bits/second. Phase quadrature modulation (PQM) can achieve four or more bits/baud to obtain 9600 or higher. Multiplexing techniques mix and match the various sources to achieve simultaneous transmission; here, frequency division multiplexing, statistical multiplexing and time division multiplexing move from frequency mixing to bit and byte interleaving in time, to dynamically interleaving transmissions that contain only the information that's present, without idle periods. This, along with various compression techniques, leads to pulse code modulation (PCM) and time space time (TST) switching, as conversations are sampled, quantized, and changed to digital bit streams, thus paving the way for the digital revolution and the end of the analog era.

Hence, over time, digital capabilities have come to replace analog in the interoffice circuits, long-distance switch centers, and now local offices. With ISDN, digital is extended to the customer premises, replacing the need for analog modems and changing the scope and range of narrowband, wideband, and broadband to what it is today. Hence, the technology is changing; the applications are changing, and the services, especially the data services, are changing and expanding, as the world becomes more and more digital.

Digital services—digital communications

With the separation of voice services from data services over the '60s and '70s, the voice network progressed to achieve a fully automated, publicly switched offering with global addresses, while data proceeded to become privately multiplexed within local area networks with local addressing, wide area bridging, metropolitan area routing, and long-distance interfacing over leased lines or satellite/microwave facilities. As time passed and more and more LANs began to proliferate here, there, and everywhere, it became essential to reconsider more integrated services and facilities in order to achieve simultaneous voice and data offerings.

In the late '70s, it became economical to digitize voice into bit streams, where 8,000 samples of the voice conversation were translated into representative 8-bit codes and multiplexed with samples of other conversations over shared facilities. In this manner, a single voice conversation requires 8 × 8,000 or 64,000 bits per second of transport facility. Twenty-four conversations are then interlaced (with appropriate network synchronization information) into a 1.544 million bits per second T1 digital communication facility. These T1s can then be multiplexed together to form T2s at 6.3 million bits/second, or T3s at 45 million bits per second. In this manner, networks were developed over the '80s to transport voice in the form of bits of ones and zeros. Transmission error rates were decreased from earlier analog achievements, as one error in 10^7 became the norm, with many digital facilities achieving one error in 10^{11} or better. This greatly improved the quality of voice transport and paved the way for its integration with data, which was also nicely represented by ones and zeros.

Beginning in the late '70s and throughout the '80s, the International Telegraph and Telephone Consulate Committee (CCITT) took up the digital services challenge. *CCITT* is a worldwide group of national Postal Telephone and Telegraph companies (PTTs) and private telecommunication companies that help set norms for input for the International Organization for Standardization (ISO) (Geneva), a worldwide association of the national standards-setting groups with the primary responsibility of creating Open Systems Interconnection (OSI) standards. Their task was to add the services to digital communications, which was proceeding nicely, as Integrated Digital Networks (IDN) were being established throughout the world. This then became Integrated Services Digital Networks (ISDN), where both voice and data services were made readily accessible to the user. ISDN used several models to accomplish this: One is the User Network Interface Model (UNI); the other is the seven-layer OSI model for denoting the stacking of communications and computer interfaces and protocols.

n-ISDN

Narrowband ISDN (n-ISDN) provides a fully digital interface that enables a single voice conversation, in digital form, at 64,000 bits/second, as well as a data-handling capability at 64,000 bits/second, with another data path at 16,000 bits per second, containing signalling/control information and a 9600 bits/sec packet-handling protocol called X.25. This basic rate interface (BRI) is defined as 2B + D, where the B channel operates at 64 Kb/s and the D at 16 Kb/s. In application, both B channels can contain a voice call or a data call. In this manner, a second voice line is available, or data can be handled at the rate of

128 Kb/s (2 × 64 Kb/s) by reverse multiplexing. In addition, eight terminals may be connected off a terminal adaptor (TA) using a collision detector algorithm and selective terminal address encoding.

In the United States, regulatory rulings don't enable the network provider (such as the Regional Bell Operating Company) to own the network terminating unit on customer premise. As a result, there's a NT2 on customer premise and NT1 on the network. While NT2 provides a four-wire 192 Kb/s interface to the terminals, special encoding enables delivery of 2B + D channels in a two-way path from the NTI to the central office at 160 Kb/s. Selective terminal adaptors provide for non-ISDN-to-ISDN interfaces for existing analog terminals based on previous RS-232, etc., -type interfaces.

ISDN uses a *seven-layer OSI model* where the first three layers—physical, data link, and network—pertain to communication needs, and the higher layers (session, presentation, and application) pertain to the computer, with the fourth layer, transport, functioning as the transition layer between the two entities.

Unfortunately, ISDN was initially marketed simply as an interface without providing access to public data networks across the country. In application, Basic Rate Interface, BRI-ISDN or n-ISDN does provide a B-channel circuit-switching data network at 64 Kb/s, 128 Kb/s, or a B-channel packet-switching data network at 64 Kb/s and 128 Kb/s, as well as the D-channel packet-switching network at 9.6 Kb/s. Primary Rate Interfaces (PRI-ISDN or P-ISDN) provides a 23B + D interface, where B is 64 Kb/s and D is 64 K b/s. These trunk-type interfaces are provided at 1.544 Mb/s, which are equivalent to T1 rates. Fractional ISDN provides N number of 64 Kb/s channels, up to the 1.544 Mb/s. Frame Relay ISDN will provide for handling LAN interfaces up to 1.544 bits/second. With proper addressing and networking features, these networks would provide a powerful universal data-handling backbone infrastructure for not only each country, but the entire world.

Broadband ISDN

B-ISDN will provide UNI interfaces of 155 million bits per second, as well as 622 Mb/s. Its transport will be based upon enveloping this information on the SONET/Synchronous Digital Network hierarchy (SDH), where SONET provides multiples of Optical Carrier (OC) rates beginning at OC-1 (50+ Mb/s), OC-3 (155 Mb/s), and OC-12 (600+ Mb/s). Broadband ISDN will use Asynchronous Transfer Mode (ATM) switching capabilities, where ATM breaks incoming information into groups of 53 bytes (8-bit) cells (containing headers), as it handles bursty, variable-bit-rate (VBR) information. Continuous-bit-rate (CBR) information can be more efficiently handled by Synchronous Transfer Mode (STM) circuit-switching techniques. It's believed that compressed HDTV will be in the form of 155 megabits or less, where videophone will be less than 50 Mb/s. Homes will be able to have up to four different video channels, with the ability to mix and match voice, data, and video to meet one's needs. As the fiber is deployed more ubiquitously, as more and more frequency wavelengths are handled, the computer-to-desktop, or home-of-the-future usage can do nothing but expand, enabling more and more services, yet to be visualized, in the Information Millennium.

Issues and observations

In many cases, 64,000-bit digital data streams are simply moving digitized analog data encoded at the slow speed of 2400 to 9600 bits/second, with corresponding parity checks for short blocks of information transfer requiring slow-speed checks, and receiving acknowledgements for each block. How can we achieve an orderly transition from the dial-up world of analog modems and multiplexors operating at low speed with slow protocols to the advanced world of high-speed digital, where we actually take advantage of using the full range of high-speed transport capabilities?

As digital replaces analog, as the Public Data Network replaces dial-up analog voice-grade data networks, what should be the deployment timeframes for achieving partial and fully available ubiquitous narrowband data services? For broadband fiber to the office? For fiber to the home?

What about standards, standards, standards? When will national ISDN be fully available for not only narrowband, but also for wideband and broadband interfaces? For wideband private-to-public data-handling networking? For videophone networking? For fully addressable service offerings?

Infrastructure

Infrastructure, infrastructure. What should it be? Who should own it? How can it be achieved? What are the tasks that lie before us? Let's begin by considering what it takes to achieve a public data switched network.

A public data switched network—thirty-two tasks

The objective is to establish within a local community by the mid '90s, an ubiquitous public data switched network offering, with general data-handling services and enhanced information service offerings for selected industries' applications. To achieve this objective, select a small group of specialists from representative marketing, network, and technical services organizations and assemble a planning team to perform the following steps.

Task one Design and determine costs for an ubiquitous narrowband ISDN "data" network for the selected area, providing the following data transport: B circuit-switched 64 K, 128 Kb/s; B packet-switched 64 Kb/s; and D packet-switched 9.6 Kb/s. Note the need to establish equipment partnerships with selected suppliers to reduce initial deployment cost, using various incentive arrangements.

Task two Define additional data-handling transport requirements: Error

Correction Detection Schemes, Store and Forward, Three Attempt Limits, Alternate Routing, Network Management, Peg Counting of Message Units, Billing, etc.

Task three Structure a dataphone address directory: Users, Closed User Group, Classes of Service, Databases, Multiple Terminals, Priority/Override, Zones, Transfers, VANs.

Task four Price data services for universal usage and growth, using several tier brackets for different ranges of usage. Resolve differences in circuit versus packet network feature packages.

Task five Establish, with selected CPE terminal equipment suppliers, offerings that can be used on a common and specific basis, such as ISDN terminals operating at 64 Kb/s, and/or specialized equipment, such as digital X-ray archive systems. Standards may need to be clarified and new standards may need to be established to ensure that proper public data network interfaces are available for internetworking operability.

Task six Establish interfaces and directories with value added networks, global VANs, and cross LATA carriers, enabling customer choice to ensure that the selected area is connected to the world.

Task seven Establish switch-to-switch interfaces to enable multiple suppliers to provide systems within the area, as well as to the value added service nodes.

Task eight Establish non-ISDN/ISDN interfaces for internetworking voice modem data to the public data network.

Task nine Design service node info switches for enhanced data transport, enabling: Delayed Delivery, Broadcast, Polling, Message Sequencing, Protocol Conversions/Code Conversions, Record Formatting, and LAN Interfacing.

Task ten Establish gateways to multiple databases, deploying menu and direct customer access.

Task eleven Establish info switch service nodes for specific applications and/or shared applications for the following communities: medical, legal, banking, and small business.

Task twelve Package offerings for mass marketing—establish packages of features and services.

Task thirteen Establish databases, either directly off the network or through gateways, through info switches or as application service centers.

Task fourteen Establish linkage networks and network directories throughout the RBOCs to Nynex, Bell Atlantic, Ameritech, U S WEST, etc.

Task fifteen Provide access to specialized software programs—third-party—enabling PCs access to shared information service centers' information handling.

Task sixteen Establish marketing organizations for the public data network on an industry basis, addressing each industry's full community of interest.

Task seventeen   Provide LAN-to-ISDN interfaces for BRI-ISDN and PRI-ISDN, as well as to frame relay-SMDS transport pipes.

Task eighteen   Establish necessary tariffs, and restructure existing tariffs to ensure ubiquitous availability and encourage increasing usage.

Task nineteen   Establish ONA interfaces to ESPs/ISPs.

Task twenty   Define a massive advertisement campaign to inform, educate, encourage, and achieve a new data user customer base.

Task twenty-one   Establish support centers for network management and data directory assistance.

Task twenty-two   Provide intercept operators to enable interfaces to selected databases.

Task twenty-three   Provide expansion plan to major cities in the state, in the region, in the nation, across the globe.

Task twenty-four   Select data organizations to establish services on the transport side, and on the service-node/info-switch side.

Task twenty-five   Provide personal training to obtain a knowledgeable work force.

Task twenty-six   Organize and use an implementation process to create and launch new services.

Task twenty-seven   Establish a network, using several equipment suppliers.

Task twenty-eight   Establish info switches with switching suppliers, and associated databases with computer firms.

Task twenty-nine   Establish info switches containing specialized software for selected industries, third-party software, specialized CPE equipment, and industry-specific software.

Task thirty   Service selected applications, as well as general applications: archive systems, database systems, ESPs, ISPs, and DBSs.

Task thirty-one   Establish state partnerships and understanding.

Task thirty-two   Launch massive advertising campaigns to bring a mountain to the community—a mountain of information creating a mountain of opportunity.

So it might be; however, congressional conditions such as the HR 5096 Bill and others continue to challenge the RBOC's right to deploy information services and ensure that the other MFJ restrictions of manufacturing and limitations to connecting LATAs remain (forever?). This will most likely drive the RBOCs to consider other approaches, such as separating basic transport from information services, as PacTel has so noted. This then brings forth a new series of issues and scenarios, depending upon the assumptions, and once implemented, a new set of risks and losses, similar to those provided by Divestiture I. So, Divestiture II may be considered in light of the issues, concerns, and opportunities, as noted by the following analysis.

The IM&M (information movement & management) company

As noted in *Global Telecommunications*, there is, and will continue to be, a market for VANs, especially as long as the public infrastructure is not fully established; but the market for VANs requires sophisticated players. Risk can be limited by concentrating capital expenditures on selected markets, but this will require an excellent plan of action, with extensive top-down planning.

Current situation

- Regulations and restrictions for providing information services continue to remain highly volatile and subject to changing decisions.
- There's a need for extensive capital to ubiquitously deploy new services while continuing to upgrade existing plant.
- The growth of voice services levels off at 2–3% per year. The growth of data services begins at 10–20% and is expected to grow extensively over the '90s.
- Private networking firms flourish, due to delays in public offerings—further removing large players from the public switched arenas.

A possibility

To meet changing regulatory boundaries, and somewhat extremely uncompromising congressional restrictions, it may be desirable to separate the Public Basic Transport Company (PBTC), which is mainly in today's voice world (POTS), and launch a fully separate entity that provides Information Switched Services (ISS) within customer premises, within the LATA, across LATAs, across the region, across the nation, and across the globe, specializing in the movement and management of information (IM&M).

This may be accomplished by: obtaining new financing from the separation of shares from the basic transport company; establishing separate revenues for a new firm dedicated to the movement and management of information—IM&M; establishing a new consortium with computer suppliers, terminal suppliers, satellite firms, national VANs, global VANs, switching/transport product manufacturers, database sources (DBS), enhanced service providers (ESP), information service providers (ISP), and large customers; segmenting the market to initially address only highly information-dependent users; deploying total information-network-handling solutions, as well as selected advanced telecommunication services by overlaying switching platforms and service platforms on the current network; addressing closed user groups for selected services, providing these offerings from shared platforms such as information switch nodes or application service centers and then extending these services to the mass market in selected areas; enabling customers direct access to these new platforms via both private and public interfaces; providing direct access to the POPs, IXCs, Global VANs, ESPs, and ISPs from these service nodes; providing services within customer premises, within the LATA, across LATAs, across the region, across the nation, and across the globe, on a common basis or for selected markets with selective offerings, such as for the local or national health-care industry.

Issues and assumptions

- No restrictions in the private deployment of the full range of information services.
- No restrictions on providing a switched information services platform that serves several LATAs.
- No restrictions in establishing private data networks across the region, nation, or globe.
- No restrictions on manipulating and processing the content of information. Code conversion, protocol conversion, data reformatting and presentation can be provided with selected third-party participants, without specialized regulatory hearings and approval.
- No pricing restraints that require specialized tariffs, other than normal business practices.
- No deployment restrictions requiring ubiquitous availability, thereby enabling selected locational offerings.
- No union personnel requirements, thereby enabling open market selection to obtain the most competent personnel.
- No extensive restrictions from the separation of the telephone operations, in inhibiting re-entrance into the local market within the former RBOC territory.
- No extensive competition from a large consortium within the territory for the first three years, as the new firm positions itself.
- No (or somewhat limited) risk in initial deployment strategies in terms of the learning curve of new management. However, past history of firms has shown that this area has the highest risk factor.
- No (or somewhat limited) success with partnerships with existing providers. Hence, the need to find a consortium of new players or go it alone.

Risks and losses

- Need for the Basic Transport Network to be upgraded in order to provide appropriate access to mass businesses and residences.
- Need for standards, standards, standards for internetworking private-to-public or private-to-private networks, especially as success in the LAN interconnect arena creates more pressure to interconnect narrowband, wideband, and broadband networks. With the lack of an aggressive basic public transport network, this might be impossible.

- Need for extensive internetworking, interprocessing, and interservices arrangements—not only with CPE terminals and systems, but with VANs, IXCs, and LECs.
- Need for low prices to encourage use and growth, causing delay in large profits, until growth occurs.
- Need to reduce initial losses, if the right technology is not available, or if marketing personnel don't use it appropriately or understand exactly what the technology can deliver, or if the reputation is damaged by improper startup problems.
- Need to provide network management, from a private arena perspective. In many cases, this is extremely complex, due to the financial limitations in establishing a large enough supporting infrastructure base for extensive throughput or emergency situations.
- Need to consider all aspects of private networking for limited customers versus pursuing combined public and private alternatives.
- Need to have the right employees, with the right training and skills, and the right products to provide the services in a timely manner.

Conclusions and observations

Even if RBOCs, such as PacTel, do divest existing telephone operations to provide advanced services from a less-regulated, separate entity, no matter which way we divide or cut the Information pie, without establishing a growing, dynamically changing, supportive information-handling infrastructure, applications can't be provided the full range of services that they require in order to appropriately meet the full range of needs across the communities (of interest) of the customers (information users). As nations like Singapore and Malaysia establish public data networks and better education for their work forces to better participate in the information marketplace, as these countries develop a prospering middle class with hope and opportunity arising from the ashes of their decayed structures, as U.S. cities in turn lose their prosperity and turn to decay and despair, as our work forces lose the American dream, as violence, crime, and immorality prevail, as higher-paying manufacturing jobs are lost to fewer, lower-paying service positions, how do we return to a competitive position of building new products with pride, having long-term financial rewards, and higher short-term profits? When shall we begin restructuring America's infrastructure? Where do we go from here? This is the challenge of the '90s for information players to resolve, as we enter the Information Millennium.

> For want of a nail, the shoe was lost;
> For want of a shoe, the horse was lost,
> For want of a horse, the battle was lost,
> For want of a battle, the kingdom was lost.
> For want of a network, the infrastructure was lost,

For want of an infrastructure, the service was lost,
For want of a service, the application was lost.

A backward glance

The telecommunications industry has now completed ten years of post-divestiture activity. It's been a somewhat turbulent and complex period—a period of positioning and politicking. With all the high expectations and excitement of the young, the new "baby Bells" have entered the marketplace and gained their new freedoms and independence. What has been accomplished over these past ten years, and how are these new companies positioned to address the market needs of the next ten years?

It's quite evident that very little substantive progress has been made in constructing a totally new "information-handling" infrastructure. The telephone company today is still "The Telephone Company." Yes, there are some new services, but for the most part, the RBOCs' businesses consist mainly of providing basic twisted-pair voice offerings. In the early '90s, capital expenditures and revenue pressures refocused the RBOCs' attention on the "telephone." After an explosion of acquisitions and mergers into every sector of business, from real estate to financial services, the majority of industry leaders remained knowledgeable in only voice-based services. This resulted in the flood of intelligent network offerings concentrating on 800 services, 900 services, the second voice line, and the number portability features of ISDN, Centrex, and Voice Mail.

Explorations into the data world generally stopped at facsimile platforms that were based upon using dial-up data modems over the analog, voice-switched network. A few companies did venture into data-packet switching, but services remained unsold due to a lack of incentive among the telco's sales force, who, for the most part, lacked even general knowledge of data users' needs. Although ISDN was devised for integrating data and voice, it was simply sold, where available, as an expensive interface for additional voice services. Few "marketeers" ever requested that the ISDN networks be deployed. For them, the network was a cloud, where data networking capabilities were ignored as data transport interoperability issues were left for the "technoids" to eventually resolve.

Many new toys were acquired along the way and then discarded once their initial sparkle became diminished, just as a child picks and chooses its favorite toy for the moment. Judge Greene attempted to put limits and boundaries on the new "babies," believing that they would completely smother anyone who attempted to join them in their newfound play box. Unfortunately, history has shown that, in truth, the RBOCs really were little threat to new competition. The RBOC's response to Judge Greene's restrictions was simply to shift their attention to other worlds and other opportunities across the sea. This resulted in little data/video switching upgrade of the American communications infrastructure, which was so badly needed to support the forthcoming information marketplace. Other competitive players, such as the information service providers, found that they needed a field upon which to play. Like it or not, it was the local operating companies who could most effectively, efficiently, and economically establish the ubiquitous arena upon which they all could play the information game. However, the RBOCs demanded relief from the MFJ restrictions before they would seriously play the game.

Without this arena, private local area networks flourished over the '80s. As they proliferated, their interconnect needs became more and more evident. As computers and communications merged, their interdependence became quite demonstrative. This demonstrated the need for internetworking, interprocessing, and interservices. In the early '90s, this apparent opportunity caused the communications industry to focus primarily on LAN-to-LAN interconnect issues, offering various forms of shared "pay as you use" transport. This causing a shift away from leased-line special services. This shift didn't go unnoticed. Alternative Transport Providers (ATPs) began offering parallel transport to encourage private-to-private networking over their privately "shared" facilities. There developed a new emphasis on providing various forms of bandwidth to the customer.

In the early '90s, there then came a mad scramble to capture the larger customers, first by dedicated facilities, and then by switched facilities. However, smaller firms realized that they needed to access larger firms' databases, and that residential services should include PC networking, database access, and video offerings. This then led to a reevaluation when video dial tone restraints were removed. Not only the FCC, but also Congress encouraged more competition and lower rates for video services to the home. Other bills, such as the Burn's bill to refiber America, gave a further push towards competition; some negative, inhibiting legislation was also enacted to reestablish information services restrictions and pull the telephone companies back to earlier boundaries. Though highly critical of the existing infrastructure, Wall Street analysts continued to require increasing short-term growth gains without providing relief for long-term infrastructure expenditures. The result was to force some of the RBOCs to consider more and more international partnerships and expenditures. They essentially became two firms: one, a local telephone company, and the other, a multinational partner in numerous international communications consortiums throughout the world.

Once removal of video dial tone restrictions was accomplished, new emphasis for CATV partnerships was promoted in an effort to quickly enter the video market using existing analog cable facilities. Later, for some RBOCs, it then became a financial choice to separate holding firms with international and CATV partnerships from the basic telephone companies. This created "Divestiture II," where the market again waited for the newborn telcos to begin structuring the missing fiber-based, local, information-handling communications infrastructure.

ISDN narrowband arrived in full force in '92 and '93, with its national ISDN—0, 1, 2, TRIP '92, and economical ISDN interface chips for CPE. This provided the impetus for medium-speed data networks to interconnect every community of interest to the marketplace. New technologies, as well as new market opportunities, challenged the industry; wireless services vied for wireline dollars. Once cellular possibilities had finally reached the acceptable price level of the $200 phone and the $30 per month rate, this network attracted not only the aggressive, overachiever executive in the BMW, but also the small-business construction worker/owner in the pickup truck. New opportunities in personal communication services arrived when the potential of the "Dick Tracy watch" captured everyone's interest. These possibilities chased scarce capital that was destined for upgrading existing plant. Similarly, satellite and microwave data networks rose to the forefront to offer wireless data transport. All of these alternatives, as well as the newfound ability to upgrade existing copper plant to handle high-speed data, challenged new fiber deployment expenditures. However, in the long run, the full spectrum of interactive fiber offerings remained awesome. Fiber In The Loop (FITL), Fiber To The Pedestal (FTTP), Fiber To The Curb (FTTC), Fiber To The Home (FTTH), Fiber To The Office (FTTO), and Fiber To The Desk (FTTD) deployments could be successfully achieved.

Therefore, there remains the need for taking the necessary steps to establish the right telecommunications infrastructure that will truly support a growing, blossoming, ubiquitous information marketplace. As we enter the next ten years after Divestiture I, the mid '90s and beyond, it's indeed a "crossroad in time" for both the computer and communications industries. It's also a "crossroad in time" for society. The right telecommunications infrastructure will enable work at home and interconnected satellite work centers in remote cities, where commuter traffic is less, homes are cheaper, children are nearer, and the quality of life is better. We're at the gate of a new frontier. It beckons us, but it offers many hazards and paths of no return. It's time to pause and consider the steps needed to successfully achieve the new information-handling infrastructure. Those who do will be better prepared to cross the new terrain and scale new heights.

Success in the global arena can only be achieved by the select few who are able to internationally hop here and there. But this is a dangerous game, as past history has adequately demonstrated. Many will venture; there will be a continuum of new partnerships and alliances as communications become more global. There's a need for financial incentives by Congress to encourage communications leaders to spend their billions locally to make new billions. To achieve the needed local information infrastructures is a 200–300 billion dollar investment by each major country. For America, this translates into a billion-dollar-per-year investment by each of the ten or so major players during the next 20 years. Then, and only then, can the destiny of the telecommunications industry be truly achieved as it merges with the computer industry.

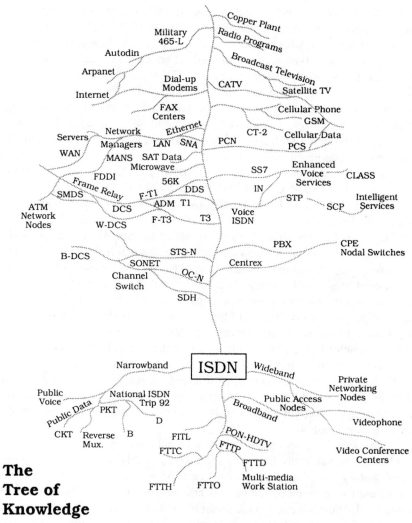

The
Tree of
Knowledge

Local-National-Global

Rural - Suburban - Metro

ATM – STM – WDM – SDM

LECs – IXCs – VANs – VARs – DBSs – ATPs

Fig. R-01. A backward glance.

A forward glimpse

"With the historic abolition of long-distance charges on 31 December 2000, every telephone call became a local one, and the human race greeted the new millennium by transforming itself into one huge, global, gossiping family."

Arthur C. Clarke
2061: Odyssey Three

We might agree or disagree with Arthur C. Clarke's intriguing vision of the world to be in 2061, but it's indeed important to assess the future of telecommunications, especially as it becomes more and more global. So, let's pause for a forward glimpse of where we're going, both technically and socially.

Let's begin by assessing the potential of future trans-switching technologies by projecting the direction and impact of SONET, fiber, photonics, switching, support, and ISDN networks. Next, let's look at future services and applications in terms of LAN, PCN, and ISDN services; then let's conclude with a glimpse of possible future cities and societies, in light of business divestitures, mergers, acquisitions, alliances and partnerships. We'll consider data networking, imaging, telephone, television, wire-line, wireless, local, and global telecommunications opportunities.

Trans-switching

Transmission and switching, previously considered two separate entities, have become more and more integrated. Variations of connectionless data transport services are routed to many different places, interfacing with connection-oriented data streams. As we look to photonic switching as the next step in the merger of transmission and switching, we see the possibility of transporting numerous wavelengths or frequencies within a single-mode fiber. As more and more frequencies (which can be visualized as colors) become simultaneously available, we can move from binary-based communications and computer systems (of ones and zeros) to other modulo-based number systems (such as decimal). In these domains, information is transported and manipulated with higher and higher degrees of content.

Just as red and blue produce violet, so multiple frequencies can produce a spectrum of colors, enabling new arrays of services. However, to both transport (switch) and manipulate information in the medium of light, we need optical memories and optical decision logic. Otherwise, we must return to the binary world of today's electronics of ones and zeros. As Dr. Ross noted, though we have prototypes of various forms of photonic switches in the laboratories, optical computer engineers are still searching for the optical equivalent of an integrated circuit. However, years of research experience have quelled the high hopes of coherent lightwave techniques boosting sensitivity one-hundred-fold over direct detection. New optical amplifiers now enable direct detection to match the performance of coherent detection. In addition, as the signal in the local area network application is combined and split into multiple channels, the optical amplifier is used in coherent detection to boost power and composite for the loss of the splitting as the original signal deteriorates. Engineers have found that Erbium-doped amplifiers, operating in the 1.5 micron wavelength, added significantly less noise from semiconductor laser amplifiers. Much more needs to be done. We need to find other elements, such as Neodymium, that can operate at an atomic emission line of 1.3 microns, as well as new technologies for transport; these might include hybrids that mix direct detection with coherent detection or lithium niobate external modulators to help reduce intrinsic noise additions and signal-loss problems. This has led researchers to believe that the commercial availability of pure photonic switches having no electronic-to-optical conversions will remain the dream product of the next century. This will leave the 2.4 Gb/s world to the '90s.

Copper plant

New technologies that enable the "mining of the copper plant" continue to indicate that expanding transport capabilities can be achieved. These could include high-speed bit rate digital subscriber line (HDSL) that uses ISDN's 2B1Q basic rate access technology to transmit 784 Kb/s over a 12,000-foot copper pair within a carrier serving area (CSA). This would enable two HDSL transceivers to send 1.544 Mb/s T1 payloads over two pairs. This, in turn, would enable users to "plug and play" with a much faster installation that doesn't require conditioning or the removal of bridge taps, which causes echoes unless removed in previous T1 installations. In addition, repeater spacing can be substantially expanded. Similarly, asymmetrical digital subscriber line (ADSL) technology links fiber and copper facilities. This provides enough capability for customers to receive VCR-quality video along with a regular telephone call. 1.5 Mb/s would be available to the customer from the network, and 64 Kb/s or 16 Kb/s would be received by the network from the customer, allowing operation "over virtually all the nonloaded loop plant up to 18,000 feet." Bellcore's HDSL Terminal Unit (HTV) is a higher-speed version of pair gain's 2B1Q transceiver for the DSL rate, where it transports 160 Kb/s for ISDN BRI over an 18,000-foot copper pair; HTV sends 800 Kb/s over a 12,000-foot copper pair—so, in parallel, two transceivers send 1.6 Mb/s over two pairs. This then extends the current plant's copper data-handling capabilities well into the next century.

Broadband

SONET (Synchronous Optical Networks), on the other hand, is like a railroad track. It doesn't care whether the train is carrying boxcars full of gold, gondola cars full of coal, flatbeds full of lumber, or all of those things together. SONET just takes them from here to there. SONET carries payloads of Asynchronous Transfer Mode (ATM) and bit streams of Synchronous Transfer Mode (STM) all mixed together. It all goes over the same fiber through the same hierarchy in a very standard way, allowing users to build different cross-sections of transport capacity and service types in the same pipe.

The future will see fiber to the curb (FTTC) and fiber to the house (FTTH) competing for implementation in active and passive optical networks (PONs) arrangements. New network strategies will take on high-speed bus and ring configurations with access switches located closer to the user; this will enable private-to-public interconnection and allow information to be: "virtually" transported through multiple survivable configurations, controlled by new superbroadband switches, and routed to information services platforms located within and outside the regulated community. As video services such as videophone are introduced, volumes of continuous-bit-rate traffic will compete with bursty, variable-bit-rate computer/terminal/workstation data traffic for bandwidth on demand. Information will be trans-switched over broadband SONET/SDH—ATM/STM—WDM/SDM facilities. Already CCITT G.707, G.708, and G.709 standards for multiple hierarchies of digital systems are accommodating the merger of North American and European differences. Multiples of the synchronous transport signal (STS-1) at 51.8 Mb/s are integrated with the European Digital Multiplexer (DSMX) 34/140 rates of 139,264 Kb/s at the Synchronous Transport Module (STM-1) rate of 155.52 Mb/s with STM-4 at 622.080 Mb/s and STM-16 at 2.48832 Gb/s.

Similarly, Bellcore TR-303 interfaces and full-access time-slot interchange (TSI) capabilities have become the centerpiece for new generations of digital loop carrier (DLC) systems. TR-303 allows a circuit interface of the DLC into the new switches, eliminating the need for additional central office (CO) terminals, main distribution frames, or other terminal gear. Backward compatibility via TR-008 enables working with existing digital carrier systems, such as AT&T's SLC 96 DLC. However, the next-generation systems will require flexible bandwidth options to enable LECs to offer inexpensive services, including n-ISDN, f-T1, T-1, DS1, T3, 384 Kb/s H/O, STS-3/OC-3, OC-12, and OC-24 (2.4 Gb/s). (See *Telephony Supercom* under references.)

As Rodney Dangerfield noted, support systems "get no respect." However, new Operating Support Systems (OSS) will be deployed over the '90s; this will enable SONET provisioning, testing, and surveillance features. Management and control will be provided over a new-language ASN.1/CMISE as a machine-to-machine language platform for exchanging messages. As the '90s progress, support systems, especially network management, customer control, and automated provisioning "get more and more respect" . . .

Finally, Open Network Interfaces, via Bellcore's TA909 specification for defining fiber in the loop interfaces, will help establish the necessary standard interfaces for both the network and the customer. Other standards are being established by T1 working groups for the Broadband ISDN and OC-1, OC-3, etc., transports via the CCITT-I series. These include the I.361 ATM layer spectrum, I.321 B-ISDN protocol reference model, and I.413 B-ISDN user network interface for permanent or semipermanent ATM virtual connection and connectionless data services. So, work continues throughout the '90s to set the trans-switching stage for the next millennium.

Wireless

In the shadows of wire-line activity, the wireless Personal Communication Network (PCN) progresses. The cellular spread-spectrum code division multiple access (CDMA) technology also vies for PCN applications to help resolve frequency overcrowding problems. CDMA is said to boost cellular capacity more than 20 times by allowing more calls to occupy the same space spread over the entire frequency band. It competes with time division multiple access (TDMA) and enhanced E-TDMA technology over the 1850-1990 MHz range. Other equipment is dispersed in the 902–928 MHz, 2.4–2.48 GHz, and 5.725-8.50 GHz radio bands to enable "the ability to make and receive a call anywhere, from anywhere in the world." This will be enabled by such future digital wireless PBX systems that have pico cells with a range of 50 to 100 meters. These provide seamless hand-over between radio cells, two-way calling and intra-building roaming.

Video

In the video arena, some believe that combining telephony with CATV will do nothing for broadband applications that will be 99 percent residential. Assuming switched video is provided over fiber, it will take 20 years to replace the metallic loop. There's little commonality between business-oriented, two-way, switched broadband service and advanced residential one-way CATV. Some look for LAN interconnection, electronic publishing, and CAD/CAM mapping rather than teleconferencing and videophone to be the global application. However, as passive optic networks are deployed for some broadband video broadcast-type services in FITL applications, CATV providers are overlaying a switched fiber arrangement on their feeder routers. They're interfacing to their traditional residential tree-branch facilities in an effort to prepare for more and more video-on-demand switched offerings. However, they're still continuing the analog world with interfaces to analog coaxial facilities to the home.

But this leads us to the future of video—not only workstations but also desktop video—as PCs begin to require more and more video communication at higher and higher resolutions. Similarly, videophone will come in narrowband, wideband, and broadband versions that require more and more transport capabilities. In time, we'll see integrated systems of varying degrees of definition/resolution as users interact. Users of the same transport quality can display similar images, while those of lesser capability require less image transport. This "mix and match" requirement will be essential during the transition and implementation period of the '90s, perhaps even throughout several hundred years of the new millennium, until sufficient areas of the globe have been upgraded sufficiently.

According to the U.S. Bureau of Census, in 1990 there were approximately 120-million workers in the United States. Approximately one-third of them worked at a desk, using PCs, dumb terminals, word processors, or one of 42 million typewriters. Fourteen million workers had PCs that used LANs or modems to interconnect them. One analysis noted that LANs were deployed such that: 75 percent of the LANs served general office applications of word processing, accounting, and database access; 8 percent education; 6 percent medical/scientific; 5 percent general; 4 percent manufacturing; and 3 percent other uses. Private LANs interconnect using Fiber Distributed Data Interface (FDDI-II) capabilities of 100 Mb/s and operating at 10 times the speed of Ethernet or token ring LANs. This was both complemented and challenged by public frame relay and SMDS-ATM switched offerings at speeds of 1.5 Mb/s to 4 Mb/s to 16 Mb/s and 140 Mb/s respectively to transport LAN data, Group 5 fax, and computer-based imagery, etc. Furthermore, narrowband ISDN (n-ISDN) supported numerous types of data communications services as well as LAN interconnect such as Group 4 fax, videotex, and computer-based imagery at medium speeds.

Thus, independent of the merger of some telcos with CATV for broadcast analog TV services, we've seen that numerous bandwidth-on-demand services, with many choices for voice/data/video offerings, can be supported by selected high-speed digital transports. These will also serve interactive two-way videophone conversations at varying degrees of compression and quality resolution until the fully digital high-definition videophone, television, and computer imagery services can enable "virtual reality" services, video "juke boxes," and image systems.

Managing the merger of C&C

As we look at the integration of computers and communications, especially as we view their mergers, acquisitions, partnerships, alliances, and divestitures, it's of the utmost importance to understand and appreciate the contrasting differences of the players that lead and manage somewhat diverse but also similar endeavors. The computer industry has been built by aggressively pursuing complex, changing, advancing technology. Aggressive management not only labored to understand the subtle differences of many new technologies, but were also willing to take the necessary risks required to achieve full potential in the marketplace.

On the other hand, advancements in the telecommunications industry were also achieved through advances in technologies. These changes were in part due to harnessing controllable segments and implementing them in five-year increments. In this manner, technology was researched extensively by laboratory personnel, carefully culled and selected, and then established in one or two new products that were then fully tested before deployment in the field.

This low-risk mode of operation was fostered and supported by a monopoly-based marketplace in which the dominant players could introduce selected new features—when and where they wanted—with little concern for the customer and without the normal competitive forces that existed in the computer arena.

The communications industry became heavily involved in ubiquitous, common, standardized offerings, while the computer industry continued to pursue, in an unrestricted manner, any new exciting idea or technological advancement. As a result, to succeed in the computer industry required superior technical skills, while to succeed in the communications industry required superior political and financial skills; the computer industry wrestled with complex technology; the telecommunications industry wrestled with massive, broad deployment plans, state and federal regulatory issues, and large organizational and personal management challenges.

As these two industries merge, especially as more C&C offerings become more and more integrated, interrelated, and interdependent, it's especially interesting to observe how the different management styles and cultures have continued to offset and inhibit each others' advancement. For example, the lack of public data network offerings drove the computer manufacturers to create their own specialized local area networks.

We should also keep in mind that the alliances between different types of communication providers also have their problems. CATV providers were able to quickly establish low-quality transport and programming offerings to get anything to the marketplace to make money. This is inconsistent with the high-quality, standardized, long-term telephone providers' mode of operation. So, as the future unfolds, as the various providers merge within the communications and computer industries, success will be tied to resolving the differences between the leaders' perspectives, modes of operations, and personal objectives.

The telecommunications industry is moving to the phase where a full evolution of the interdependent parts must take place. As we move into the age of light in the forthcoming Information Millennium, fiber optics will play an essential role as light is split, multiplexed, demultiplexed, and switched. As noted, laboratories throughout the world are hard at work on devices such as multifrequency lasers and the optical equivalent of electronic transistors, amplifiers, dividers, and switches. The telecommunication potential of direct optical signal processing and switching is enormous. Realizing this potential might require some redirection of efforts or midcourse corrections to better channel the forces of change in the direction we want to take.

The Information Millenium

The early 20th century fostered rapid growth in the telephone industry, as numerous separate telephone companies were established in the cities. In the final years of this century, we've seen the expansion of private local area networks (LANs) searching for interconnectivity, with their terminals and computers searching for interoperability. Hopefully, the future possibilities of new public/private switching networks will offer an insight into the proper integration of computers and communication so that the appropriate public data narrowband, wideband, and broadband networks can be established in a timely, universal manner; this will diminish the unneeded chaos and complexity of attempting to integrate numerous dissimilar networks without the supporting common communications infrastructure.

To finish our assessment, we need to pause and take a final look at society in America and across the globe. We have indeed come to a "crossroad in time." Just as past technologies challenged the horse and buggy world of our ancestors to advance to the fast-paced, machine-based society of today's high-powered sports cars and jet airplanes, so will future technologies have similar, far-reaching effects as we advance to a world based upon the creation and use of information. In the past, industrial nations of the world have found their growth in removing farm workers from the harness of the plow to the harness of the assembly line. Both blue-collar and white-collar workers migrated to the factory and office jobs of the cities. This extensive growth formed metropolises, where large labor pools moved from one production line to another. This enabled Los Angeles' military industry, Seattle's aerospace industry, Boston's computer mainframe industry, and Chicago's electronic television/communication industries to flourish and blossom. Subsequent economic forces caused migration of these industries to every sector of the globe in search of cheaper and cheaper labor.

With the advent of the integrated circuit and its subsequent computer in a chip, a new age had dawned. A shift occurred from hardware assembly lines (that employed many uneducated people) to more automated assemblies that used sophisticated equipment to speed the flow of products and improve their quality. Today, computers have entered every aspect of the marketplace; their versatile programming capabilities have changed every form of work. As a result, previous repetitive or complex tasks are replaced by computer-controlled systems. No longer are large labor pools required in centralized locations; more distributed but more educated forces provide the different pieces of the product.

As the wild twenties spawned new jobs, good times, excessive income, and questionable morality, so the wild eighties enabled the haves (separated from the have nots) to experience prosperity, along with considerable immorality; this has left us with the "morning after" residue of problems to clean up and resolve in the nineties. Large corporations have begun to refocus on their basic businesses, following years of improper expenditures, reduced products and production lines, closed plants, and high debt. Workers have experienced dwindling jobs, dwindling salaries, and dwindling opportunity. Unfortunately, this boom-to-bust scenario is not simply the result of the classic business cycle of expansion and recession; several key shifts and changes have taken place.

As those within the communications industry know, technology is now able to redistribute the urban work forces back to the rural communities, where smaller groups of workers are networked together by new electronic/ optical information highways to form "virtual" assembly lines across the region, nation, or globe; these "assembly lines" will produce new products from new ideas. With the renewed emphasis on customer needs and new products to fit specific user applications, there will be an exciting array of new services over the next millennium. This will be achieved in a new, competitive, global marketplace. Countries such as Singapore, Malaysia, Hong Kong, Korea, and Japan are leading the way in establishing versatile communications infrastructures to participate in the information revolution. These and other aggressive nations are addressing the many complex issues of today as they attack poverty and despair by establishing new jobs, providing education for the masses, winning the war on drugs, and encompassing a return to family values and morality. Without addressing these needs, once-prosperous countries will diminish, while others will grow stronger. We've seen the rise and fall of many civilizations over the ages. Hopefully, these new technologies can provide some assistance as we take our first steps on the long journey back to a vibrant, competitive, strong, and morally sound society. So, lets close with a look into the future. It may well be a tale of two cities.

A tale of two cities

Infolopolis and Megatropolis are two urban cities located in different states, somewhere in America. Megatropolis is a thriving, hustling, bustling community built on power; here, the decision-makers gather, clustering their headquarters together to influence government policies and procurements. Many of their plants are nearby; others are selectively located in appropriate congressional districts, where sympathetic congressman that concur with their interests prevail. In a different time, in a different place, Infolopolis has risen from the ashes of a decaying city, where new technologies and global competition brought a quick and forceful blow to the city's major industrial plants. However, after a long delay, visionary city and state leadership brought an immediate reversal to a degenerating society, bringing new hope and prosperity to its inhabitants. In order to better appreciate the ramifications of the forthcoming information marketplace, let's review the day-to-day activities of these two entities, Megatropolis and Infolopolis.

Megatropolis Megatropolis is built on the East Coast, in a corridor of large cities. Most of the inhabitants of Megatropolis are within and around an inner core of skyscrapers. The local communications monopoly had been under the stringent control of the local telephone company, which specialized in automating voice operating systems and deploying accessible LAN interconnect transport. The telephone company has entered the television transport in joint partnership with local CATV providers. It didn't offer specialized public networks to facilitate universal data or interactive video; it believed that it was the task of the customers to construct their own networks, using shared bandwidth, public transports; these were needed to offset the high costs of privately leased transport. Alternative networks were offered in the wireless domain by constructing an overlay cellular network; in the wire-line arena, there was a joint offering with the local cable company to enable selective access to "video juke boxes" for pay-per- view services. Due to the limited technical needs, little fiber, if any, was deployed in the residential plant for this endeavor; the telephone company instead relied mainly on CATV's coaxial analog cable transport for delivery, using specialized, selected frequencies for low-speed customer requests or voice conversations. Remote data customers were encouraged to use dial-up analog data modems to access the information services providers' databases or gateways to LANs. The main data effort consisted of interconnecting LANs and providing service platforms that provided various protocol and code conversions to internetwork various types of LAN transport.

It was interesting to note that, as time progressed, several alternative "networking" service providers were established to provide similar protocol interfacing software packages. In time, many of the major businesses elected to privately internetwork their own traffic via gateway switches on their premises, using the public network only as an overload safety valve. In this manner, life remained relatively the same within the city.

However, as daily functions began to turn to using the telephone network to transport more and more dial-up data via analog modems, traffic "brown outs" began to appear during busy hours. The analog voice-based switching systems could not keep up with the excessive number of attempts or the holding times required for low-speed data traffic. As time passed, city-wide data networks began to appear. These selectively removed the various industry-based communities of interest from the public network. The local government chose to interconnect their offices by using LANs within their various complexes, and by establishing large, survivable, ring-type transports using *dark fiber* (fiber in conduit without public network electronics). They purchased this from the local telephone firm, but added their own transmission electronics with connections to internal nodal switches. Selected locations with specialized nodal switches did provide access to the world and allowed direct connection to a particular interexchange carrier's point of presence. They also facilitated limited outside access to their internal facilities via dial-up modems.

In this manner, the city businesses huddled together to create as much efficiency as possible in their use of their private internetworking facilities. Non-electronic, face-to-face meetings continued to be the best form of communication, and this encouraged local automobile transportation and global airplane transportation. As more time passed, congestion from local traffic growth increased; the city complex grew and grew until gruesome. Soon, it became more economically productive to establish private microwave links between the office complexes surrounding the city and the downtown headquarters; this enabled video conferencing of meetings to cut down on local travel.

However, problems continued to prevail as cross-communities of interest attempted to interconnect information. For example, doctors in one HMO had difficulty sharing information with specialists or doctors in another HMO. Similarly, lawyers had difficulty accessing insurance firms and state law enforcement files because each was self-contained on their own internal networks. As time progressed, addressing problems severely hampered cross-network assessability. Layers of protocols were required to achieve interconnection. Much was accomplished in this regard, but the complexity and congestion factors kept the networks at the edge, leaving users many times more willing to use their automobile transportation rather than telecommunications. In time, living in Megatropolis became more and more complex when efforts to offset automobile congestion encouraged the work force to live in the more densely packed high-rises, rather than suffer the long commute times caused by too much rush-hour traffic. As the businesses within Megatropolis became more and more successful, their success generated more and more growth, which generated more and more communication needs that could not be fulfilled by their existing local telephone infrastructure, highway transportation infrastructure, or social community infrastructure. This caused many to question the quality of life in Megatropolis.

Infolopolis On the other hand, the city leaders of Infolopolis were keenly aware that a new revolution was taking place. Their large assembly plants were using somewhat unskilled labor and attempting to compete with foreign firms that were using electronically controlled mechanical devices for automated assembly. As the highly competitive Japanese plants led the way in robotic assembly and "just in time" inventory control, manufacturing facilities all over the world became more and more automated. The leaders of infolopolis indeed realized that a new revolution was taking place—an information revolution. To be part of this change, to harness it and channel it for their own use and advancement, they elected to construct a new communications infrastructure to support new businesses that used information as one of their key tools and assets. Their first step was to expand existing facilities by constructing a parallel network to the voice network—a public data network over the existing copper plant. To do so, they used the full capabilities of narrowband ISDN. Every home and business within the entire city, as well as each major town in the surrounding communities, were provided access to the new public data network. Soon, new public "dataphone" directories were established so that every terminal was interconnected. Security and password codes were developed to protect closed user groups from outside penetration. Care was taken to ensure that the new network was equipped sufficiently.

New, shared work centers were established to ring the city with electronically interconnected facilities; this encouraged workers to live nearby and commute to work locally, rather than over long distances. These centers were excellent candidates for the faster information transport speeds provided by first wideband and then broadband, publicly switched and addressable facilities. New, higher-speed switched public networks were established to enable extensive computer-to-computer traffic, video imagery, multinode worksta-

A Telephonic
Telecommunications Beginning

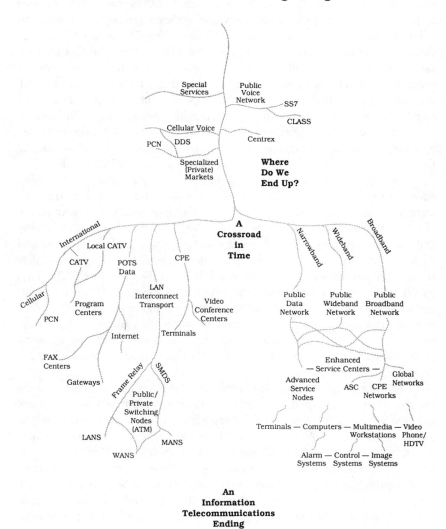

Fig. R-02. A forward glimpse.

tions, video conferencing, and high-resolution videophone capabilities; this more fully integrated the remote locations together as one. Access was made to national and global networks with special, reduced tariffs to encourage use.

After the initial years of growth, an interesting phenomenon began to take place. During the fifth year and the years following, new cities began to spring up; they were quite remote from the initial city. Huge communication highways were constructed to link these cities together, as new, more distant rings of cities began to cluster around the city center. But these rings didn't develop as an urban/suburban extension. They had sufficient green belts (miles and miles) between them to ensure that small-town lifestyles were achieved, but they had electronic access to all the services of a large city. In this manner, satellite work groups could develop their own particular parts of a particular product, which was then electronically integrated into the total package. Hardware assembly plants did exist, using the latest in automation, but these plants could now be located in more remote communities, because of "online" access to design information from other remote locations. So, as Infolopolis developed over time, many different community-of-interest groups were established using the common public ISDN network as the basic transport infrastructure.

Specialized information service platforms were then constructed to tailor offerings to specific needs. These platforms were established above the network in a layered networks' layered services architecture. Cross "community of interest" interoperability was achieved using public addressing, switching, and standard open interfaces. As broadband fiber was publicly switched and deployed to every residence, interactive videophone services augmented data services to provide full multimedia capabilities to the new users. The local telephone company became the local information company. Numerous information service providers were now able to provide a growing array of services to meet growing customer needs. Private networks did exist within local complexes, but they used the public network for interconnection, as private-to-public internetworking became quite standard and straightforward using addressable, switched offerings.

Cellular and private alternative networks also developed selectively as the massive transport capabilities of the fiber delivered more and more massive data and video offerings, successfully providing a parallel, economical alternative to coaxial-based CATV networks. Much was spent on upgrading the local infrastructure, for to go "global" was to create a robust "local" marketplace. It was back to rebuilding the "business of the business." In time, these increasingly high-quality communications were not only available for selective local applications, but were also interconnected to the world through global communications, which as Arthur Clarke noted, were to be as relatively inexpensive as a local phone call.

It was a time of growth. It was a time of improving the quality of life for the citizens of Infolopolis (a distributed INFOrmation-based, LOcal megatroPOLIS). (Figures R-0 through R-19 are provided for your reference and to demonstrate the full buildup from customer needs to network services.)

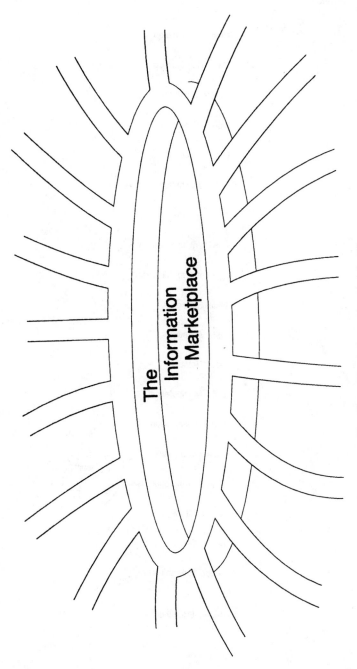

The Information Marketplace

Fig. R-0. The information marketplace.

Timely Information

Time Management

Queue Elimination

Inquiry/Response

Data Collection

Data Distribution

Data Transfer

  • Point to Point

  • Point to Multipoint

Data Storage

Data Access

Data Retrieval

Data Manipulation

Data Presentation

Delayed Delivery

Image Generation

Image Storage

Video Display

Video Entertainment

Video Education

. . . .

. . . .

Fig. R-1. Customer needs.

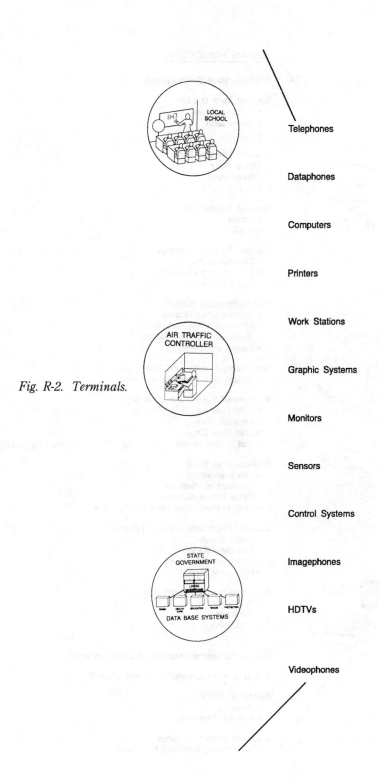

Fig. R-2. Terminals.

Telephones

Dataphones

Computers

Printers

Work Stations

Graphic Systems

Monitors

Sensors

Control Systems

Imagephones

HDTVs

Videophones

GENERAL APPLICATIONS

RES/BUS Voice Communication

Business Point of Sale
- Hardware Store
- Supermarket
- . . . .

Business Inventory Control
- Parts Store
- Burger King
- . . . .

Business Transactions
- Orders
- Buy/Sell

Business Financial Exchange
- Currency Exchanges
- Credit Card Exchanges
- Check Exchanges

Medical/Education Records
- Patient/Student Records
- Examination/Test Records
- Diagnostic/Recommendations
- Course of Program

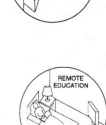

Legal/Library/Event/Topic/...Cross-Index
- Multiple Data Base Access/Storage

Automobile/T.V./Furniture/...Assembly
- Parts Flow
- Assembly Flow
- Lead Time Order
- Just In Time Arrival

Fig. R-3. General applications.

Business Data Transport
- LAN Interconnect
- Computer to Computer
- Virtual Private Networking
- Point to Point Video Conf. Center (private)

Education/Government Private Networking
- Image Transfer
- Bulk Data Transfer
- Dedicated Video Links
- Point to Multi-point

Residential/Business Videophone
- Narrowband
- Wideband
- Broadband

Business/Government/Manufacturing Workstations

Buisiness Video Conference Centers (Shared)

Residential HDTV
- Broadcast
- Selected Programs

Global Information Exchange
- Narrowband, Wideband, Broadband

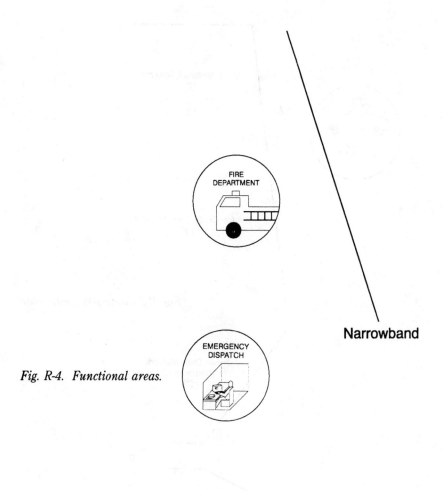

Fig. R-4. Functional areas.

Narrowband

Wideband

Broadband

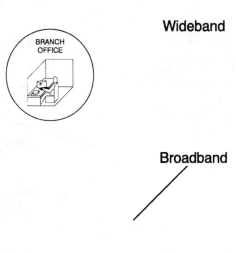

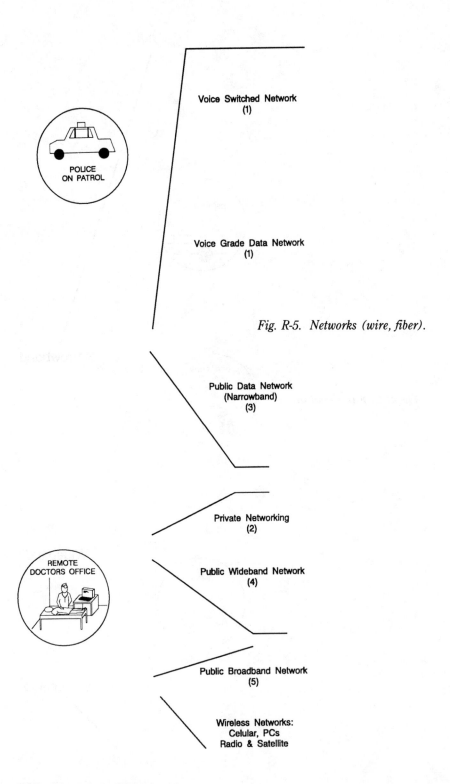

Voice Switched Network
(1)

Voice Grade Data Network
(1)

Fig. R-5. Networks (wire, fiber).

Public Data Network
(Narrowband)
(3)

Private Networking
(2)

Public Wideband Network
(4)

Public Broadband Network
(5)

Wireless Networks:
Celular, PCs
Radio & Satellite

POLICE
ON PATROL

REMOTE
DOCTORS OFFICE

Telephone
Dial tone
IXC Carrier of Choice
Custom Calling
800/900
Class
Weather/Time
Centrex/Centron
Voice Mail
Inward Dialing (PBX)
IXC Carrier of Choice

Data Sets
Dial Up Fax

Fig. R-6. Services.

Dataphone
Data Dial tone
E-Mail
Broadcast
Polling
Delayed Delivery
Protocol Conversion   C
Code Conversion
Record Reformatting
Blockage
Secure/Private
Info Switch Access
Videophone I

Imagephone
Customer Bandwidth Selection
   • By Order
   • Dynamically
Customer Network Selection
   • By Order
   • V.P.N.   D
Survivability
Priority
Secure/Private
VideophoneII

Videophone IIII
Video Dial tone
Workstation (Multimedia)
HDTV
Image Storage┐ Delayed┐ Random┐ Search
Video Storage┘ Delivery┘ Access ┘ Browse
Computer to Computer Transparency
Gateway Menu
Narrowband/Wideband/Service Integration
   • C
   • D

4 Khz-Voice

— — — — — — — — — — — —

4Khz-Data
(Non-conditioned facilities)
• 2400 b/s
• 4800 b/s
• 9600 b/s
(Conditioned facilities)
• 14.1K b/s
• 19.2K b/s
• 56K b/s

— — — — — — — — — — — —

2B+D
2x64K b/s+16K b/s ≤ 160K b/s ≤ 192K
b/s
64K, 128K b/s CKT
64K, 128K b/s PKT
   9.6K b/s PKT

BRI-
I
S
D
N

Fig. R-7. Transport.

23B+D
23x64K b/s+64K b/s=1.544M b/s
NX64K b/s, 1.544M b/s (f-T1)
1.544M b/s (T1)
1.544M b/s
NX64K b/s ≤ 1.544M b/s (f-ISDN)
45M b/s (T3)
MX 1.544M b/s (f-T3)
MX64K b/s ≤ T3

PRI-
I
S
D
N

51M b/s
155M b/s
620M b/s
1.2M b/s
2.4M b/s
4.8M b/s

B-
I
S
D
N

LARGE
BUSINESS

Dial tone
IXC Carrier of Access

– – – – – – – – –

Dial Up Data
@2400 b/s - 1 in 10⁴ Error Rate
Point to Point - Conditioned (C1, C2)
Availability 99.5 to 99.9%
Switched 56K

– – – – – – – – –

Fig. R-8. Features.

Data Addressing
Switched 64K CKT, PKT
Switched 9.6K PKT
Store & Forward
Error Rate 1 in 10^7 A
   • Error Correction/Detection
Alt Routing Survivability
IXC Data Carrier Access

Data Addressing
Frame Relay Interface   P
FDDI Interface   R
Ethernet Interface   I
                          V
Token Ring/Bus Interface A   P
SMDS Interface   T   U
Connection Less   E   B   B
Connection Oriented   L
Switched/Non Switched   I
ISDN/Non ISDN   C
POP Access
P&P Inter-networking

Switched Videophone
Switched Multimedia Workstation
Point to Multipoint HDTV
Global Interconnection
Info Switch Access
Narrowband/Wideband/Transport
Integration
   • A
   • B

SPECIALIZED
DATA BASE

LARGE MAINFRAME

WEATHER CONDITIONS
DISASTER RECOVERY
NETWORKS

Timely Information
Time Management
Queue Elimination
Inquiry/Response
Data Collection
Data Distribution
Data Transfer
  • Point to Point
  • Point to Multipoint
Data Storage
Data Access
Data Retrieval
Data Manipulation
Data Presentation
Delayed Delivery
Image Generation
Image Storage
Video Display
Video Entertainment
Video Education

Fig. R-9. Customer needs.

Fig. R-10. Customer needs—terminals.

EMERGENCY
DISPATCH

AIR TRAFFIC
CONTROLLER

BRANCH
OFFICE

CUSTOMER NEEDS   TERMINALS

Timely Information	Telephones
Time Management	
Queue Elimination	Dataphones
Inquiry/Response	Computers
Data Collection	
Data Distribution	Printers
Data Transfer	
• Point to Point	Work Stations
• Point to Multipoint	
Data Storage	Graphic Systems
Data Access	
Data Retrieval	Monitors
Data Manipulation	Sensors
Data Presentation	
Delayed Delivery	Control Systems
Image Generation	Imagephones
Image Storage	
Video Display	HDTVs
Video Entertainment	
Video Education	Videophones

Fig. R-11. Customer needs—terminal-general applications.

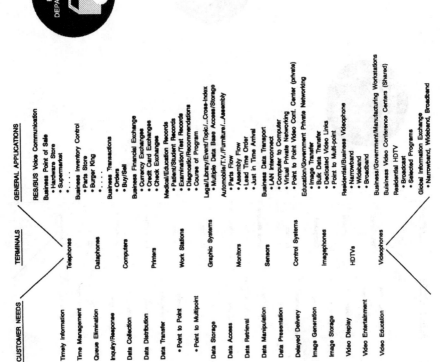

CUSTOMER NEEDS

- Timely Information
- Time Management
- Queue Elimination
- Inquiry/Response
- Data Collection
- Data Distribution
- Data Transfer
  - Point to Point
  - Point to Multipoint
- Data Storage
- Data Access
- Data Retrieval
- Data Manipulation
- Data Presentation
- Delayed Delivery
- Image Generation
- Image Storage
- Video Display
- Video Entertainment
- Video Education

TERMINALS

- Telephones
- Dataphones
- Computers
- Printers
- Work Stations
- Graphic Systems
- Monitors
- Sensors
- Control Systems
- Imagephones
- HDTVs
- Videophones

GENERAL APPLICATIONS

- RES/BUS Voice Communication
- Business Point of Sale
  - Hardware Store
  - Supermarket
- Business Inventory Control
  - Parts Store
  - Burger King
- Business Transactions
  - Orders
  - Buy/Sell
- Business Financial Exchange
  - Currency Exchange
  - Credit Card Exchanges
  - Check Exchange
- Medical/Education Records
  - Patient/Student Records
  - Examination/Test Records
  - Diagnostic/Recommendations
  - Course of Program
- Legal/Library/Event/Topic/...Cross-Index
  - Multiple Data Base Access/Storage
- Automobile/T.V./Furniture/...Assembly
  - Parts Flow
  - Assembly Flow
  - Lead Time Order
  - Just in Time Arrival
- Business Data Transport
  - LAN Interconnect
  - Computer to Computer
  - Virtual Private Networking
  - Point to Point Video Conf. Center (private)
- Education/Government Private Networking
  - Image Transfer
  - Bulk Data Transfer
  - Dedicated Video Links
  - Point to Multi-point
- Residential/Business Videophone
  - Narrowband
  - Wideband
  - Broadband
- Business/Government/Manufacturing Workstations
- Business Video Conference Centers (Shared)
- Residential HDTV
  - Broadcast
  - Selected Programs
- Global Information Exchange
  - Narrowband, Wideband, Broadband

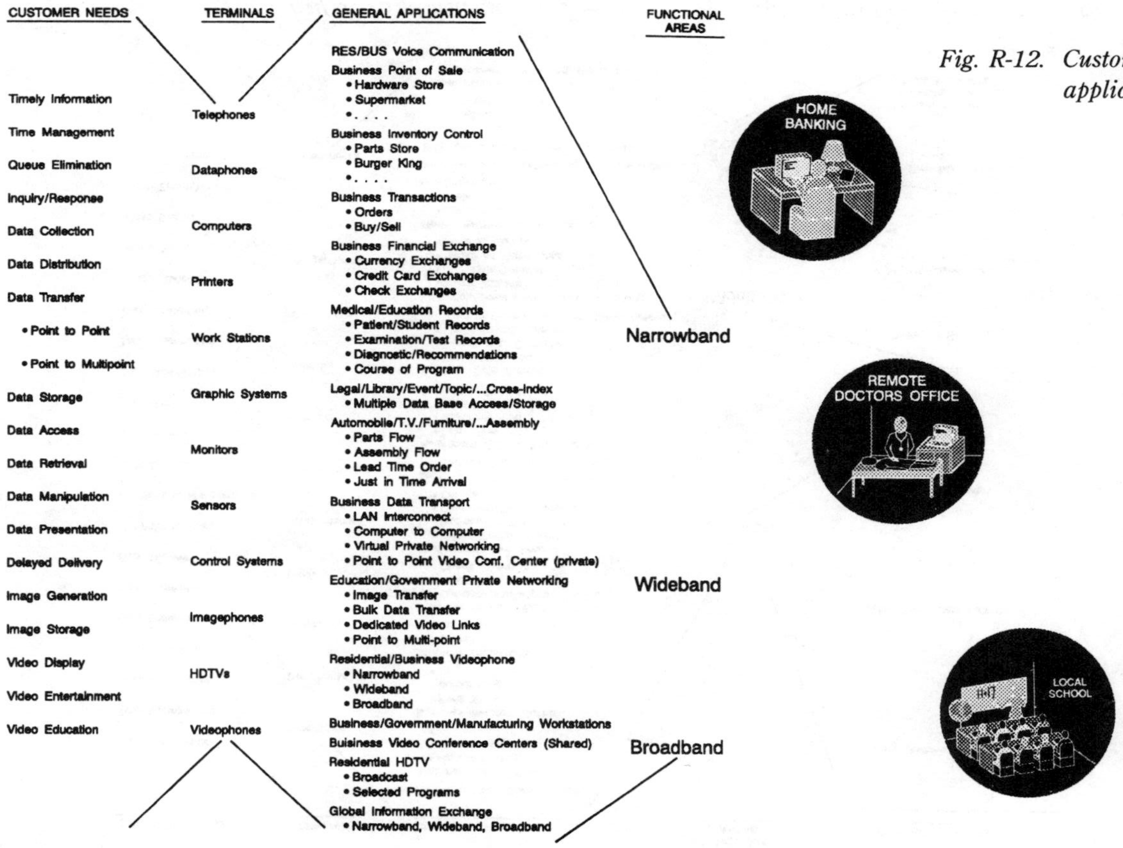

Fig. R-12. *Customer needs—terminal-general applications, functional areas.*

FUNCTIONAL AREAS

HOME BANKING

REMOTE DOCTORS OFFICE

LOCAL SCHOOL

Narrowband

Wideband

Broadband

CUSTOMER NEEDS	TERMINALS	GENERAL APPLICATIONS
Timely Information	Telephones	RES/BUS Voice Communication
		Business Point of Sale
		• Hardware Store
		• Supermarket
Time Management	Dataphones	Business Inventory Control
		• Parts Store
Queue Elimination		• Burger King
Inquiry/Response	Computers	Business Transactions
		• Orders
Data Collection		• Buy/Sell
Data Distribution	Printers	Business Financial Exchange
		• Currency Exchange
Data Transfer		• Credit Card Exchanges
		• Check Exchanges
• Point to Point	Work Stations	Medical/Education Records
		• Patient/Student Records
• Point to Multipoint		• Examination/Test Records
		• Diagnostic/Recommendations
		• Course of Program
Data Storage	Graphic Systems	Legal/Library/Event/Topic/...Cross-Index
		• Multiple Data Base Access/Storage
		Automobile/T.V./Furniture/...Assembly
Data Access	Monitors	• Parts Flow
Data Retrieval		• Assembly Flow
		• Lead Time Order
		• Just in Time Arrival
Data Manipulation	Sensors	Business Data Transport
		• LAN Interconnect
Data Presentation		• Computer to Computer
		• Virtual Private Networking
Delayed Delivery	Control Systems	• Point to Point Video Conf. Center (private)
		Education/Government Private Networking
Image Generation	Imagephones	• Image Transfer
		• Bulk Data Transfer
Image Storage		• Dedicated Video Links
		• Point to Multi-point
Video Display	HDTVs	Residential/Business Videophone
		• Narrowband
Video Entertainment		• Wideband
		• Broadband
Video Education	Videophones	Business/Government/Manufacturing Workstations
		Business Video Conference Centers (Shared)
		Residential HDTV
		• Broadcast
		• Selected Programs
		Global Information Exchange
		• Narrowband, Wideband, Broadband

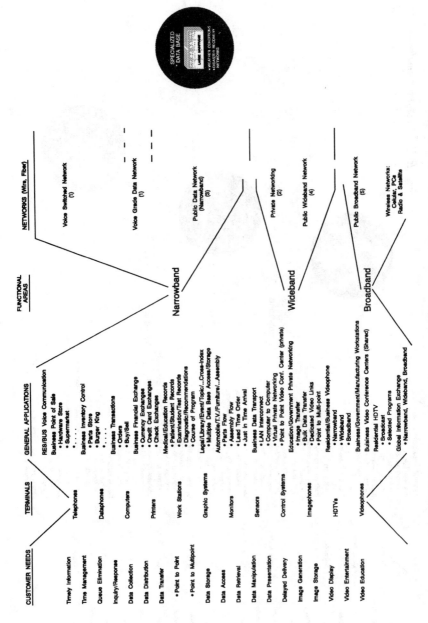

Fig. R-13. Customer needs—terminal-general applications, functional areas, networks.

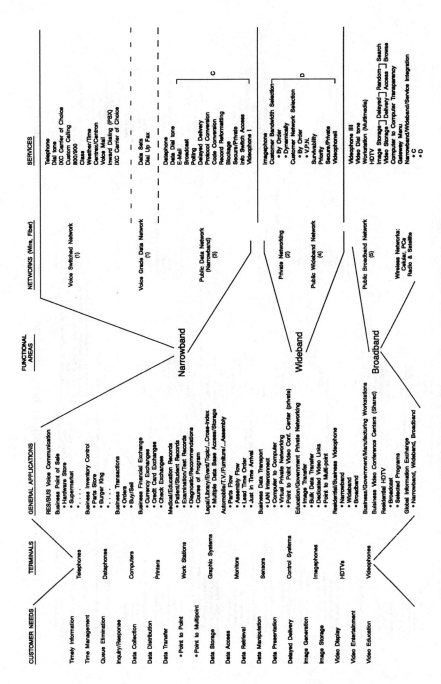

Fig. R-14. Customer needs—terminal-general applications, functional areas, networks, services.

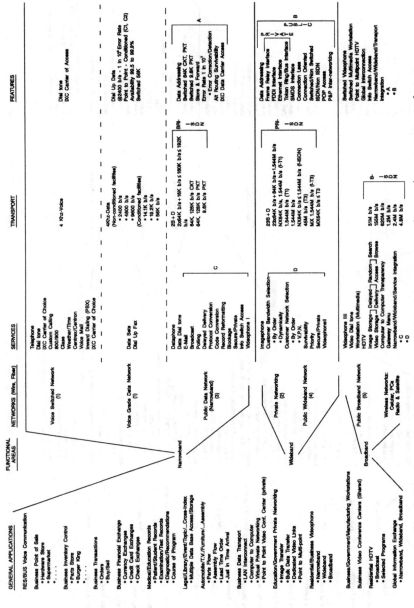

Fig. R-15. General applications, functional areas, networks, services, transport, features.

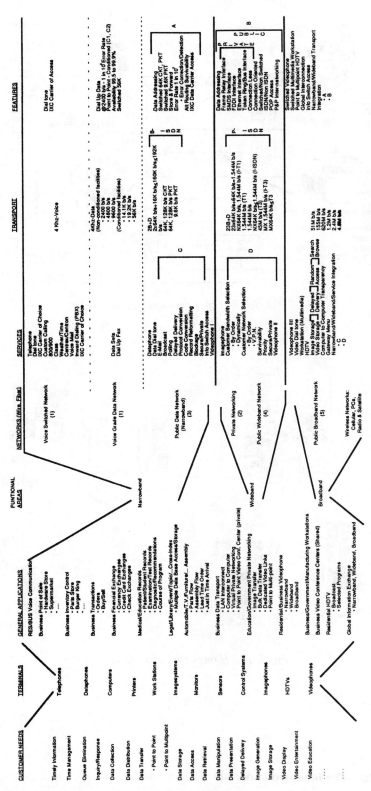

Fig. R-16. Customer Needs—terminal-general applications, functional areas, networks, services, transport, features.

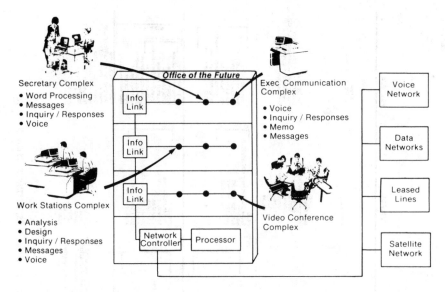

Fig. R-17. Office communications center.

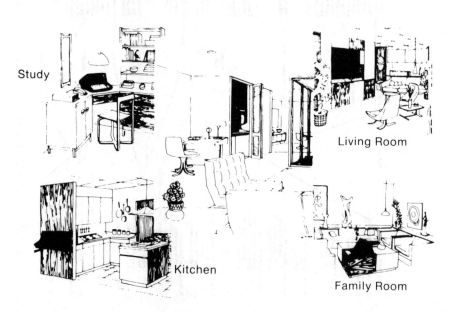

Fig. R-18. Home communications center.

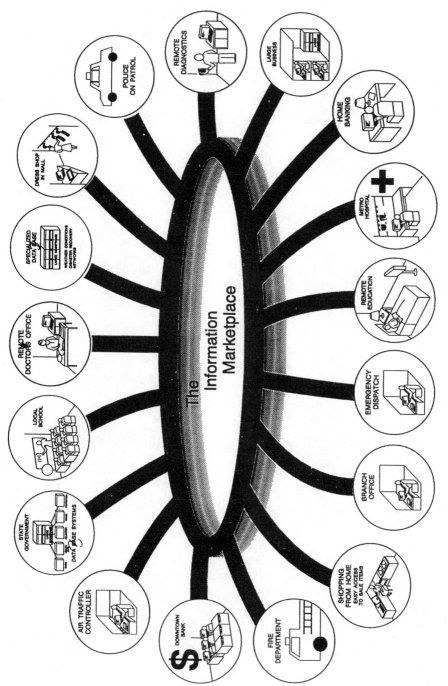

Fig. R-19. *The Information Society.*

References

AIN public policy issue statements. Bellcore to National Services Coordinating Group, November 13, 1991.

Bellcore NOINCC Docket #91-346. "Intelligent networks." January 23, 1992.

Burke, John. "The day the universe changed." 1990-1991 TV series.

Encyclopedia Britannica, 1992.

FCC Docket #91-346. "The matter of intelligent networks notice of inquiry." December 6, 1991 summary.

Heldman, Robert K. 1992. *Global telecommunications*: Layered networks' layered services. Blue Ridge Summit, Pa.: TAB/McGraw Hill.

———. 1989. *ISDN in the information marketplace*. Blue Ridge Summit, Pa.: TAB/McGraw Hill.

———. 1987. *Telecommunications management planning: ISDN networks, products and services*. Blue Ridge Summit, Pa.: TAB/McGraw Hill.

History of mathematics. IBM, 1966.

Telephony Supercom Review, 1992.

World book encyclopedia, 1992.

Glossary

A wise man once noted to his son, who later told his son, that 75% of any new job (be it engineering, marketing, or whatever) is simply understanding (and remembering) its new terminology. With this in mind . . .

Bits and bytes Information can be coded in the binary form of ones and zeros as bits, or grouped in 8-bit bytes to denote alphanumeric characters or 16 to 24-bit pixels of information.

Community of interest Indicates the span of access and the interaction of multiple users within a functional area, such as the medical industry. Here, the flow of information exchange can be observed by flow charts denoting all forms and types of information handling across the marketplace, addressing the broad picture and denoting the various parties' role in the creation, access, storage, and manipulation of information.

Applications Only when technology is applied to real marketplace applications, addressing and resolving customer needs, will the feature on the feature and/or the service on the service become apparent. Herein lies the key to success. Too often, if the initial trial does not provide immediate success, the technology is aborted. Only in continuously pursuing numerous applications can the right technology, with the right twist, the right turn, be identified.

Features and services Products provide features of this or that capability, which then need to be packaged into a desirable service that truly meets customer needs.

Industry Within each market sector, there are one or more industries that have both common and/or separate communication/information-handling needs. Some industries span several market sectors, while others are subsets of industries within a single sector.

Information handling Inquiry/response (I/R), data collection (DC), data distribution (DD), transactions (T), remote documentation/documentation (RD/D), enhanced voice (EV), and videophone/video conference (V/

VC) functions are performed for the movement and management of information (IM&M).

Information marketplace The descriptive term to indicate the new marketplace formulated by the shift from purchasing physical products manufactured on assembly lines, established by the Industrial Revolution, to the market denoted by the creation, transport, storage, access, retrieval, manipulation, and presentation of voice, data, text, image, graphic, and video information.

Information networks Narrowband, wideband, and broadband transmission-switching capabilities enable information services to be transported here and there locally, nationally, and internationally, via networks. While narrowband refers to the movement of information up to, and including, two 64 Kb/s channels, wideband provides for multiple 64 Kb/s channels up to 1.5 Mb/s and 6.3 Mb/s over copper facilities, as well as for 45 Mb/s throughputs over fiber. Broadband transport capabilities are based upon the Synchronous Optical Network/Synchronous Digital Hierarchy (SONET/SDH) structure, beginning with the Optical Carrier Rate-One (OC-1) of 50+ Mb/s and on to OC-2, 3, etc., in multiples of 50+ Mb/s to as high as Gigabit OC-24, OC-48, etc., rates. Broadband will absorb both narrowband and wideband as lower rates are packaged within Asynchronous Transfer Mode (ATM) and Synchronous Transfer Mode (STM) variable and continuous rate bit streams.

Information services Voice, data, text, image, graphic, and video services enable information to be created, accessed, browsed, indexed, stored, retrieved, transported, manipulated, collected, condensed, compressed, listed, cross referenced, packaged, and presented in individual or multimedia forms, over narrowband, wideband, or broadband facilities.

Information switching Circuit, packet, channel, cell, fast packet, burst, ATM, STM, Hybrid. As information is transported from point to point, and point to multipoint, there are several types of systems for selecting, switching, and routing the varying bit streams. In circuit switching, paths are made available prior to the call on a dedicated basis or dynamically at the time of the call, and held for the duration of the call. In this manner, virtual private networks (VPNs) can be dynamically established and changed to this or that destination. In packet switching, the message is broken into segments and transported in packages, having headers and tails, to various destinations that either remember the movement of previous packets (connection-oriented) or individually handle each packet without prior destination knowledge (connectionless) movement. When information is broken into exceedingly small segments called cells, these cell streams can be switched in bursty or fast packet-handling arrangements where information arrives asynchronously and is transferred across the switch to synchronous SONET transport networks. In the asynchronous mode of operation (ATM), varying information can be asynchronously packaged in 48 fixed eight-bit/one-byte cells, with five bytes of overhead. While, in channel switches, information is transported as a variable number of channels of 64 Kb/s; the Synchronous Transfer Modes (STM) switches handle a fixed number of channels, such as at the 1.5 Mb/s, 45 Mb/s, or 155 Mb/s rate.

Information transport   LANs, WANs, MANs, Frame Relay, SMDS, FDDI, SONET, SDH. As information is transported internally within a local area network (LAN) on buses or rings to various terminal devices that are selectively addressed as a private network, it can then be routed over publicly switched facilities to more remote locations on wide area networks (WANs) or on private metropolitan area networks (MANs) for closed user groups. Here, internetwork routing and addressing is performed to enable LAN-to-LAN accessibility. In time, publicly available, dynamically changeable transport facilities can be shared using SMDS capabilities over SONET transports providing for a variable number of frames of information to be relayed from point to point, and providing access to privately shared higher-speed transport FDDI facilities.

ISDN/INIS/LNLS   Integrated Services Digital Network, discussed in *Telecommunications Management Planning*, provides a parallel public data network offering that must enable Integrated Networks' Integrated Services operations, as noted in *ISDN in the Information Marketplace*, which are structured on a Layered Networks' Layered Services infrastructure, as presented in *Global Telecommunications*.

Market sector   To best segment and differentiate consumer needs, separate markets can be addressed by focusing on selected areas, such as: business, education, residence, government, as well as subsections, denoted as: large and small business, state and federal government, etc. In this manner, further analysis can denote geographical differences, demographical concentrations, financial constraints, etc.

ONA/AIN/INA   Open Network Architecture, Advanced Intelligent Network, and Information Network Architecture are all interrelated. INA and AIN enable multiple providers, both network and service, to be accessed by multiple users from both private and public networks. This indeed is the challenge of a functioning open network. However, this network must be capable of survival, while maintaining integrity and privacy.

P&P   Private and public internetworking is the game of the '90s and beyond—and "tying it all together" is the objective of the new information game.

Products and platforms   New switching systems can be established in the public network to enable access and transport routing to traditional common carrier central offices (COs) or alternative carriers' point of presence (POPs). Here, several new Class-level systems can be provided closer to the user, thereby extending the traditional five-level network switching hierarchy to the customer premise by deploying Class 6-level access nodes and internal Class 7-level customer switches. Enhanced services providers (ESPs) or information services providers (ISPs), as well as the various carriers, can provide advanced services by locating, above the network, service nodes that provide switching, specialized information handling, and access to common databases or specific access to specialized databases in application service centers.

S&S Securable, integratable, private, and survivable transport is becoming increasingly important as users change from using dedicated, leased-lines special services to shared switched public offerings, requiring 100% availability, security, and survivability. Here, integrity and privacy must be guaranteed, for not only transport, but for database access.

Terminals With terminals, the telephone makes way for the dataphone and then to the videophone, where personal computers, workstations, and knowledge stations access and manipulate information; where picturephone, videophone, viewphone, video conferences, and video workstations portray individuals using various forms of compression and resolution algorithms; while television advances to ATV and then to high-definition (HDTV) capabilities as users are provided increasingly less expensive and more economical narrowband, wideband, and broadband transport capabilities.

User types The type of information user for a specific application can be identified in terms of the type of information handling performed on a specific task, requiring specific information network capabilities. The users can then be viewed on a specific industry basis, or across industries within or across market sectors to denote commonality, as well as individually required information services.

Universal information services marketplace FCC decisions in the Fall of '92 promoted full competition within the local arena and challenged the RBOCs' local monopoly by authorizing direct connection throughout the LATA to alternative transport providers (ATPs), competitive alternative providers (CAPs), competitive access providers (CAPs), value added networks (VANs), value added resellers (VARs), information service providers (ISPs), and enhanced service providers (ESPs). Peter Huber's Geodesic Network began to take shape and substance. However, without participating RBOCs' narrowband, wideband, and broadband networks, this ruling leads to new shapes and forms that even Peter had not visualized, leaving us with more questions than answers as to who are the "keepers of the networks." Who will control and administer addresses for not only the voice networks but also the data and video networks? Who will ensure that Open Network Architecture (ONA) internetworking protocols and standard interfaces are achieved? Will alternative provider's equivalent Class 5s, 6s, and 7s be given blocks of area/office (10 digit/7 digit NNX) codes or ISDN (15 digit) codes? Will collocation or "virtual" collocation be achievable for not only trunk access but shared dialtone? Will multiple services from multiple providers, using intelligent networking call-processing interrupts, severely affect the network, especially during busy hours? When will the layered networks' layered services approach be embraced and utilized as the model to achieve the competitive, open information marketplace visualized by the FCC? These are the questions for the '90s, as more and more services such as video dialtone by the RBOCs, regulations on CATV basic rates, and interactive data networks blossom to bring the world closer and closer together.

Index